Introduction to Synthetic Aperture Radar

About the Author

E. David Jansing, Ph.D., is a chief engineer at the Johns Hopkins University Applied Physics Laboratory and a lecturer in the Johns Hopkins University Whiting School of Engineering's Engineering for Professionals program. He has been studying synthetic aperture radar for more than 20 years and has been teaching a graduate-level course on SAR for nearly 10 years. Dr. Jansing has more than 20 years' experience in remote sensing, especially in imaging, exploitation, analysis, and data collection/planning for SAR, hyperspectral imaging, and infrared imaging. Machine learning, particularly deep learning, has been the focus of his exploitation efforts and he has a patent for a novel maritime vessel detection using synthetic aperture radar. A graduate of the University of Louisville, Dr. Jansing has spent time working the research, development, and operational sides of military, intelligence, and commercial remote sensing. He currently lives in Cathedral City, CA.

Introduction to Synthetic Aperture Radar

Concepts and Practice

E. David Jansing, Ph.D.
Johns Hopkins University Applied Physics Laboratory

Mc Graw Hill

New York Chicago San Francisco Athens
London Madrid Mexico City Milan
New Delhi Singapore Sydney Toronto

Library of Congress Control Number: 2021931309

Introduction to Synthetic Aperture Radar: Concepts and Practice

1 2 3 4 5 6 7 8 9 CCD 26 25 24 23 22 21

ISBN 978-1-260-45896-1
MHID 1-260-45896-2

Sponsoring Editor
Lara Zoble

Editorial Supervisor
Stephen M. Smith

Production Supervisor
Pamela A. Pelton

Acquisitions Coordinator
Elizabeth M. Houde

Project Manager
Warishree Pant,
KnowledgeWorks Global Ltd.

Copy Editor
Tulsi Rawat,
KnowledgeWorks Global Ltd.

Proofreader
Raghu Narayan

Indexer
Arc Indexing, Inc.

Art Director, Cover
Jeff Weeks

Composition
KnowledgeWorks Global Ltd.

Contents

Preface

If you are new to synthetic aperture radar (SAR), you should be warned: SAR is just complicated. There is no other way to put it. A friend of mine put it this way: "SAR theory is fractured." And it is. There is little to no standardization in vocabulary, variables, or techniques.

There are so many different ways to look at SAR: systems engineering, radar engineering, signal processing, algorithm development, automatic target recognition, statistical processing. I doubt there is a text that sufficiently covers all of these areas simultaneously and certainly this book doesn't try. This book approaches SAR from a signal processing perspective, as many SAR books do. However, this book is unique in the sense that it explicitly commingles imaging geometries and breaks out the mathematical models by demodulation type (quadrature demodulation versus dechirp processing). It is the demodulation type that typically drives image formation processing, with the image geometry (such as stripmap versus spotlight) playing a secondary role (or no role at all, in some cases). (A notable exception to this is wide-aperture SAR, where the need to fully compensate range cell migration for image formation is likely the driving factor.)

This book evolved from years of working in and teaching synthetic aperture radar. There are several excellent texts that examine SAR from a specific vantage point, often geometry or system driven. When teaching, it's difficult to ask students to purchase multiple books for a single course. For the practitioner, it is more convenient to have a single reference for an overview, rather than have five or six SAR theory books on your shelf. This work pulls together all of the materials needed for a graduate-level course in synthetic aperture radar, and presents the material in a fashion that will be useful for the seasoned practitioner as well.

The reader should come to this topic with a solid background in digital signal processing. If terms such as Nyquist sampling rates, Fast Fourier Transforms, and windowing are foreign to you, you might consider building a foundation in signal processing first before tackling this topic. The reader should also have at least a basic understanding of radar before approaching this topic. While there is some discussion of the radar range equation and antenna beam patterns, this work idealizes most of the radar engineering portion of SAR. It is assumed that the power levels, analog-to-digital conversion, and local oscillators are perfect (or at least very well behaved). There are plenty of sources of error, but for an introductory text, it seemed like overkill to identify them all.

The reader should approach this book in the order it was written, which is from foundation to application. For the novice, the first few chapters that cover the geometric and signal models are essential and cannot be skimmed. Full understanding of later chapters, including image formation, SAR applications, and image quality techniques, requires that mathematical foundation. Novices must read and understand the first four chapters before moving to more advanced topics. Practitioners looking for an easy reference to advanced topics (autofocus, for example) are welcome to skip ahead, realizing that the lack of standardization within the SAR communities may cause some confusion. A variables list and glossary are included at the end of the book for easy reference. A conscious effort was made to standardize variables within the text itself, so the variables presented here may not match the variables as they were originally published in their respective journal or conference articles, or as they appear in other texts.

Finally, there was a conscious decision made to not include full blocks of code in this book. Programming languages evolve rapidly and it isn't clear to anyone what development language will be in popular use in 10 years' time. At the time of this writing, MATLAB from MathWorks and open-source Python with data analysis packages are two popular development and data analysis languages in use among SAR practitioners. R, however, is gaining traction as a well-rounded statistical and algorithm platform. Including all of these choices in entire blocks of code is a daunting (and perhaps impossible) task. So, if there is any code at all, it is small snippets of code in MATLAB and Python that hint to implementation of specific steps (such as the proper way to instantiate a Fast Fourier Transform), not entire blocks of code that accomplish a specific algorithmic task, such as the Polar Formatting Algorithm (PFA). It is my considered opinion, having wrestled with SAR and taught SAR to others for decades, that the best way to understand a specific concept in SAR is to code it up yourself. There are open-source toolboxes available for various development language flavors, including MATLAB and Python. Notable contributions in this area are the National Geospatial-Intelligence Agency's MATLAB SAR Toolbox[1] and Python SAR Toolbox, SARPy,[2] and RITSAR,[3] a Python tool set developed by a student from the Rochester Institute of Technology. If algorithm development is not your strong suit, you might consider one of these toolboxes for implementation and understanding of various topics covered in this book.

E. David Jansing, Ph.D.

[1] https://github.com/ngageoint/MATLAB_SAR

[2] https://github.com/ngageoint/sarpy

[3] https://github.com/dm6718/RITSAR

Acknowledgments

It all started with a single image. While working on my dissertation at the University of Louisville, my advisor and friend Dr. Darrell Chenoweth handed me a single synthetic aperture radar to study. That image and research led me to Naval Air Weapons Station, China Lake (NAWS/China Lake), in Ridgecrest, CA, where I spent a fascinating and educational summer in 1996 studying SAR and various ways to use it. Mr. John Knecht served as my mentor there and I'm extremely grateful to him and Dr. Chenoweth for my introduction to the world of SAR.

My introduction to the elegant and challenging details of SAR processing came from the late Dr. Stuart Melzer while I worked at The Aerospace Corporation. Stu was a patient teacher who, despite my struggles in understanding, took his time to show me the way. His influence over me in this topic was powerful and I will be forever in his debt. I'm also thankful for the interactions and technical discussions I had with my Aerospace colleagues, particularly in my development of short courses in SAR. The genesis of this material started there.

My work in SAR was further refined when I moved to the Johns Hopkins University Applied Physics Laboratory. It was there that I was invited to produce a full-semester, graduate-level course in the Johns Hopkins University Engineering for Professionals program, which is part of the Whiting School of Engineering. I am eternally grateful to Dr. Brian Jennison, the chair of the Electrical and Computer Engineering Department where the SAR course lives, for his gentle and patient encouragement in the development of the course and this text. I'm thankful to the Janney 2.0 program within the Applied Physics Laboratory, which partially funded the writing of this book, giving me a good deal of freedom to work on this text during my work day. Their financial contribution to this book cannot be understated; without it, I fear that the text would never have been written.

I am also extremely grateful to my APL colleagues Mr. Chad Hawes, Mr. Howard Schoeberlein, Dr. Chung-Fu Chang, Dr. Mark Baker, Mr. Chris Watkins, Dr. Rick Chapman, Dr. Steve Scorpio, Dr. Greg Avicola, Dr. Michael Nord, Dr. Eric Ericson, Dr. David Porter, Mr. Eric Larsen, Dr. Chris Gifford, Dr. Rich Gasparovich, Dr. Christopher Boswell, and Dr. Bob Henrick for their time, patience, and understanding in helping me refine my knowledge of SAR through discussion and application. I'm the sort of person who relishes discussion and conversation among a large number of folks. It is, in essence, how I learn. It typically involves a whiteboard (in the office) and sometimes a bourbon (outside the office). Thanks to all of you for providing such a supportive and educational working environment. And thanks to my previous and current supervisors,

Ms. Cyndi Utterback and Ms. Anna Slowikowski. Their encouragement and support played a significant role in the creation of this book.

And to my editor, Ms. Lara Zoble at McGraw Hill, thank you for your patience through this process. Without your guidance and support, this text would not be possible. I want to extend my thanks to my students who helped edit the text before it was sent to the publisher, as well as to my intern, Ms. Danielle Hardy, who provided valuable comments during the writing phase of the book.

I am indebted to both Riley Hansen and Larry Moore at ImSAR. They provided system specifications for their airborne SAR platforms, called NanoSAR Pods (NSPs), as well as example imagery from those pods that they kindly granted me permission to reprint. Their fine imagery is displayed on the front cover of this book. I've had the pleasure of working with Riley and Larry performing data collections "in the field." Their willingness to help, to advocate, and to provide cannot be understated. I am also grateful for imagery and support provided by ICEYE. Their beautiful imagery also adorns this text. I would also like to thank Capella Space Systems for their contributions of imagery and system specifications to this book. They are new to the SAR world and their recent successes produce marvelous, beautiful, and useful imagery.

I am compelled to thank my family for their patience, support, encouragement, and drive. My parents, Barbara and Bob, planted the idea in my head that I could do anything, if I just tried hard enough. I have worked hard not to prove them wrong and I am eternally in their debt for encouraging that philosophy. In 1981, they purchased my first computer, a Commodore 64. It started a love affair with computers and algorithm development that burns hot in me to this day. My sister, Rebecca, is an educator of fine young women and men in the sciences. Her undying devotion to her students is a shining light and a source of enlightenment and encouragement to me. I thank her and her husband, Andrew, who suffered through engineering school with me at the University of Louisville, he as a civil engineering student and I as an electrical and computer engineering student. And to their adult children, Ian and Elaine, thanks for letting me be the cool yet nerdy uncle. To the Walters—Granny, Grumpy, Ed, Erin, Cassie, Patrick, Tyler, Ronan, Kira, and Megan—thank you for all of the support, encouragement, and shared meals where the deeply dangerous question was asked, "So what is it that you do, exactly?"

To Larry and John, thank you from the bottom of my heart. Your support and encouragement were sometimes the only things keeping me going.

Finally, to my grandparents, Edna, Robert Sr., Angela, and Michael, thank you. No words could fully describe the positive and loving impact you had on my life and my work. May all of you rest in peace. This work is fully dedicated to your memory.

E. D. J.

CHAPTER 1

Introduction to Synthetic Aperture Radar

In this chapter, the very basics of radar will be presented, along with the advantages to radar imaging over passive optical imaging and some examples of existing synthetic aperture radar (SAR) sensors. Additionally, the basic geometry model will be presented, along with some of the terminology that is unique to SAR. Finally, the term "synthetic aperture" will be defined, as well as why it is needed to produce high-resolution radar images.

1.1 Introduction

Synthetic aperture radar[1]—typically abbreviated SAR, which can cause confusion with the acronym for "search and rescue"—is a strange beast. Few understand it's value. To most, it's a poor substitution for an optical image.

Figure 1.1 shows a comparison of an optical image, shown at the top of the figure, with a SAR image, shown on the bottom of the figure.[2] The SAR image appears "grainy" or noisy as compared to the optical image. The SAR image is from TerraSAR-X, a German SAR satellite. Both images have a ground sample spacing of 1.25 meters. The concept of resolution in SAR is somewhat different than that of optical imaging; this will be covered later in the book.

This comparison is a bit misleading. Optical sensors measure incoming radiation in either the solar or thermal regions, making them *passive*. SAR sensors measure incoming radiation in the radio wave region; the incoming radiation is provided by either the SAR sensor or another (sometimes noncooperative) radiation source, making SAR an *active* sensing modality. This makes SAR a "24/7" technology, meaning, it can be used any time of the day or night. It can also see through most types of weather, including clouds and rain (this is wavelength dependent). NASA's Magellan spacecraft is an excellent example. Launched in 1989, it traveled to Venus with a SAR and mapped 98% of Venus'

---

[1] "RADAR" itself is an acronym for RAdio Detection And Ranging or RAdio Direction And Ranging. People rarely treat it as an acronym anymore. This text will use "radar" instead of "RADAR."

[2] N. Merkle, W. Luo, S. Auer, R. Müller, and R. Urtasun, "Exploiting Deep Matching and SAR Data for the Geo-Localization Accuracy Improvement of Optical Satellite Images," *Remote Sensing*, Vol. 9, No. 6, p. 586, June 2017, doi: 10.3390/rs9060586.

FIGURE 1.1 A comparison between an optical image (top) and a SAR image (bottom). [*Image credit: Merkle, Luo, Auer, Müller, and Urtasun (2017). Reprinted under CC BY-SA 3.0.*]

surface at resolution of about 100 to 150 meters, an impossibility using optical imagery with the extremely cloudy atmosphere that shrouds Venus.[3] Figure 1.2 shows the SAR of Venus, as taken by Magellan.

Optical sensors are typically used to measure an object's reflectivity or emissivity.[4] These are the qualities that give way to "color" in the visible light region of the

---

[3] https://solarsystem.nasa.gov/missions/magellan/in-depth/

[4] R. A. Schowengerdt, *Remote Sensing: Models and Methods for Image Processing*. Elsevier, Amsterdam, 2007.

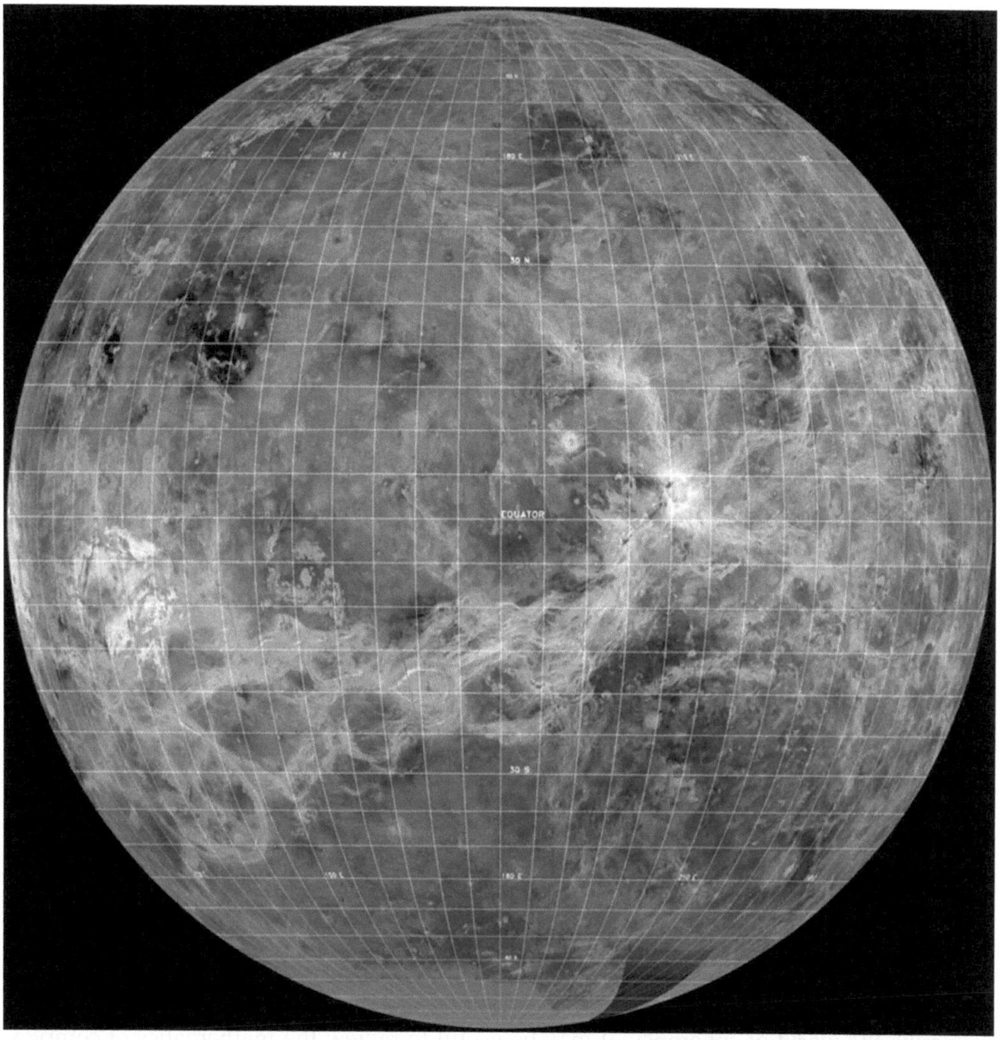

FIGURE 1.2 A synthetic aperture radar image of Venus from NASA's Magellan spacecraft. (*Image credit: NASA/JPL.*)

electromagnetic spectrum. SAR, on the other hand, is used to measure radar cross-section.[5] Cross-section is analogous to reflectivity or emissivity, in that it acts as a multiplier to the return signal's amplitude. In the case of SAR, however, this is a *complex* multiplier, not a scalar one as it is for optical imaging.

SAR is a *coherent* imaging modality. That is, phase is preserved in the imaging process. Optical imaging, on the other hand, is typically *incoherent* (there are some notable

---

[5] M. Skolnik, *Radar Handbook*, Third Edition. McGraw-Hill Education, New York, NY, 2008.

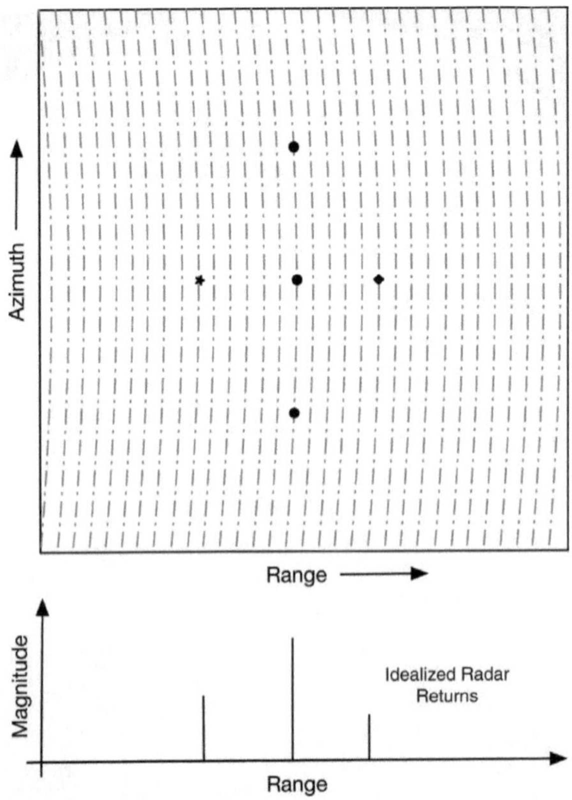

FIGURE 1.3 Several point scatterers placed on iso-range lines in image scene.

exceptions, like optical holography). As we will see much later in the book, this coherency leads us to several extremely useful by-products, like interferometry, tomography, moving object detection, and coherent change detection.[6] Complex numbers describe the world of coherency; it may be useful to review the mathematics of complex numbers.

To understand the history of SAR, it's important to understand why SAR is needed. Radar is first and foremost a *ranging* device. An object's distance can be determined by measuring the round-trip delay time it takes for a transmitted waveform to get from the antenna to the object and back. We call that object a *scatterer* or a *target* (with a nod to radar's military history[7]). Since the travel time is directly proportional to the speed in which the wave traveled, which is at the speed of light, the distance is easily calculable. See, for example, Figure 1.3, which shows five targets in a hypothesized patch on the

[6] Very simply: complex numbers are used to describe rotation, which is just phase for sinusoidal waveforms.

[7] C. Hulsmeyer, "Hertzian-wave projecting and receiving apparatus adopted to indicate or give warning of the presence of a metallic body such as a ship or a train," British Patent No. 13170, 1904.

ground. *Range* refers to the direction away from the radar antenna (sometimes mistakenly called the "cross-track" direction, but it is only truly cross-track when the antenna is pointed *broadside* or at an angle perpendicular from the flight track). *Azimuth* refers to direction perpendicular to range. Again, when the antenna is pointed broadside, azimuth is "along-track" and range is "cross-track." In the figure, the gray dotted lines represent lines of constant range (also called iso-range lines). The star-shaped scatterer, which is closest to the radar, is on the left, the circle-shaped scatterers are along an iso-range near the center of the scene, and the diamond-shaped target is on the right. A non-imaging radar would be able to distinguish the star-shaped scatterer and the diamond-shaped scatterer with no difficulty as their return times will be different; the star-shaped scatterer would return before the diamond-shaped scatterer. The radar would also be able to distinguish that there is at least one target near the center of the scene. These idealized returns are shown in the plot below the image scene. Since the three circle-shaped scatterers are all on the same iso-range line, there is no way for the radar to separate them. Imaging radars seek to distinguish targets distributed on a patch of ground within some resolution cell. In early SAR systems, the resolution cell was large, on the order of hundreds of meters to kilometers. But as systems became more sophisticated and as digital technology became readily available, the resolution cell shrank to sub-meter levels. As Cumming and Wong write in their 2005 book,[8] "There is no need to explain the virtues of digital technology here. It is fair to say that a large proportion of innovations in radar systems in the last 30 years has arisen from the use of digital technology in radar systems design, especially in data processing."

1.2 Origins of SAR

Both Curlander and McDonough[9] and Carrara, Goodman, and Majewski[10] have excellent histories on the origin of radar and SAR. These histories will be highlighted here.

Since the early part of the twentieth century, radar had been used by the military to protect assets and to seek out targets of interest. Hulsmeyer demonstrated in 1903 that a radar could be used for ship collision avoidance; he later patented the idea.[11] In the 1940s, the proximity fuse was developed in England,[12,13] and refined by the Johns Hopkins University Applied Physics Laboratory.[14] The refinements included miniaturizing the small, short range, Doppler radar. The small radar sensed the proximity of the target

[8] I. G. Cumming and F. H. Wong, *Digital Processing of Synthetic Aperture Radar Data: Algorithms and Implementation*. Artech House, Boston, MA, 2005.

[9] J. C. Curlander and R. N. McDonough, *Synthetic Aperture Radar: Systems and Signal Processing*. John Wiley & Sons, New York, NY, 1991.

[10] W. G. Carrara, R. S. Goodman, and R. M. Majewski, *Spotlight Synthetic Aperture Radar: Signal Processing Algorithms*. Artech House, Boston, MA, 1995.

[11] C. Hulsmeyer, "Hertzian-wave projecting and receiving apparatus adopted to indicate or give warning of the presence of a metallic body such as a ship or a train," British Patent No. 13170, 1904.

[12] J. P. Baxter III, *Scientists Against Time*. MIT Press, Cambridge, MA, 1946.

[13] J. W. Brennan, "The Proximity Fuse: Whose Brainchild?," *United States Naval Institute Proceedings*, Vol. 94, No. 9, 1968.

[14] L. Brown, "The Proximity Fuse," *IEEE AES Systems Magazine*, July 1993.

FIGURE 1.4 The Willow Run stripmap synthetic aperture radar image, taken in August 1957. (*Image credit: Willow Run Laboratory, University of Michigan. Reprinted under CC BY-SA 3.0.*)

and detonated, rather than relying on a direct hit, which was much more difficult with the highly maneuverable aircraft of World War II.

While there were several organizations investigating the use of a "synthetic" aperture for distinguishing targets in both azimuth and range, Carl Wiley of Goodyear Aircraft Corporation is generally credited with the creation of SAR.[15] Wiley noted that two different fixed targets could be resolved separately using Fourier or frequency analysis. The angular separation relative to the motion of the aircraft provides the means for a sort of Doppler that could be measured to some fidelity, that fidelity being dependent on the radar bandwidth (in range) and the angle between the start and end of the illumination period, often called the *dwell*. Wiley called his technique Doppler beam sharpening. The Goodyear group built the first SAR airborne system in 1953 and flew it on a DC-3.[16] The system operated in L-band at 930 MHz and achieved a synthetic beamwidth of approximately 1°. This equated to an azimuth resolution of about 10 meters. However, it is unclear when the first focused SAR image was actually taken. Curlander and McDonough claim that Goodyear built the first SAR airborne system, while Carrara, Goodman, and Majewski state that the first successfully focused airborne SAR image was taken at Willow Run airport in August 1957.[17] These confusions and contradictions may stem from the covert nature under which SAR was developed; details of these secret systems may not have been revealed until many years or decades later. This image is shown in Figure 1.4.

Another group at the University of Illinois also experimented with SAR. This group was able to image a seven-mile-long stretch of Key West in Florida in 1953.[18] The signal processing technique used by the team was only able to produce an image well-focused in azimuth over a small range interval.

---

[15] C. A. Wiley, "Synthetic Aperture Radars: A Paradigm for Technology Evolution," *IEEE Transactions on Aerospace Electronic Systems*, Vol. AES-21, 1985.

[16] J. C. Curlander and R. N. McDonough, *Synthetic Aperture Radar: Systems and Signal Processing*. John Wiley & Sons, New York, NY, 1991.

[17] W. G. Carrara, R. S. Goodman, and R. M. Majewski, *Spotlight Synthetic Aperture Radar: Signal Processing Algorithms*. Artech House, Boston, MA, 1995.

[18] C. W. Sherwin, J. P. Ruina, and R. D. Rawcliffe, "Some Early Developments in Synthetic Aperture Radar Systems," *IRE Transactions on Military Electronics*, Vol. MIL-6, No. 2, 1962.

The pioneers of SAR in the 1950s and 1960s paved the way for SAR systems that could take high-resolution images from both aircraft and spacecraft, over Earth and planets beyond. Here are some notable, but nowhere near exhaustive, examples of operational SAR systems.

1.3 Examples of Synthetic Aperture Radar Systems

There are several dozen SAR systems, past, present, and planned, that could be outlined here. This section will focus on five Magellan, TerraSAR-X, Capella Space Systems, ICEYE, and ImSAR.

1.3.1 Magellan

The concept of mapping the surface of Venus using radar was conceived by NASA scientists in the late 1970s.[19] That first mission was to be called the *Venus Orbiting Imaging Radar* (VOIR), but the program was cancelled in 1982.

A new mission was proposed in 1983 as the *Venus Radar Mapper*, which was a simplified radar mission, composed of four main objectives: (1) Collect radar images of the surface of Venus with a 1.0-kilometer resolution, (2) create a topographic map with 50 kilometers spatial and 100 meters vertical resolutions, (3) collect gravity field data with a resolution of 700 kilometers and an accuracy of 2 to 3 milligals, and (4) derive an understanding of the geological structure of Venus.

The radar system was ultimately named RDRS. The spacecraft, shown in Figure 1.5, arrived at Venus on August 10, 1990 and mapped 98% of the planet in mapping missions that lasted until September 14, 1992.[20] The satellite was decommissioned on October 11, 1994 when it was commanded to plunge into the planet's dense atmosphere. The purpose of this maneuver was to understand the planet's atmosphere and characterize the performance of the craft as it descended.

The key radar characteristics include a SAR that operated at 2.385 GHz (S-band), a peak power of 325 watts, a pulse length of 26.5 μs, a pulse repetition frequency (PRF) of 4.4 to 5.8 kHz, a swath width of 25 kilometers, a data acquisition rate of 806 kbps, and a downlink quantization of 2 bits. This is worth noting that with only 2 bits per sample, NASA was able to create high-resolution, high-contrast images of the Venusian surface. This is because SAR maps each unique point in a scene to a two-dimensional (2D) sinusoid corresponding to that point. This will be explained in more detail in the next chapters. Thus, the important information is the frequency and phase of the sinusoids for each point in the scene, which is fundamentally different than optical imagery, where the sampling of illumination in a chosen color space dictates the fidelity and contrast of a given aerial or satellite image.

---

[19] Wikipedia, "Magellan (spacecraft)," `https://en.wikipedia.org/wiki/Magellan_(spacecraft)`, 2019.

[20] N. Aeronautics, S. A. (NASA), and J. P. L. (JPL), "Magellan Mission to Venus," `https://www2.jpl.nasa.gov/magellan/`, 2019.

FIGURE 1.5 An artist's depiction of the Magellan spacecraft. (*Image credit: Wikipedia/Wikicommons. Reprinted under CC BY-SA 3.0.*)

1.3.2 TerraSAR-X

Launched on June 15, 2007, TerraSAR-X was Germany's first operational radar satellite.[21] A joint venture between the German Aerospace Center (DLR) and EADS Astrium GmbH (since acquired by Airbus), TerraSAR-X is a high-resolution SAR satellite that provides high-quality radar data commercially. Additionally, along with its sister satellite TanDEM-X, the two satellites are able to use bistatic radar and interferometry to create high-resolution digital elevation map (DEM) products.[22,23]

The system is capable of collecting high-spatial and radiometric resolutions. TerraSAR-X can collect single, dual, and quad polarizations, as well as multitemporal imaging data. Interferometric products, including repeat-pass and along-track interferometry, are also available.

TerraSAR-X was considered such a technological success that, in 2010, an entire issue of IEEE Transactions on Geoscience and Remote Sensing was dedicated to the system

[21] R. Werninghaus and S. Buckreuss, "The TerraSAR-X Mission and System Design," *IEEE Transactions on Geoscience and Remote Sensing*, Vol. 48, No. 2, pp. 606–614, January 2010.

[22] eoPortal Directory, "TSX (terrasar-x) mission," `https://directory.eoportal.org/web/eoportal/satellite-missions/t/terrasar-x`, 2019.

[23] eoPortal Directory, "TDX (terrasar-x) mission," `https://directory.eoportal.org/web/eoportal/satellite-missions/t/tandem-x`, 2019.

(IEEE Transactions on Geoscience and Remote Sensing, Vol 48, No. 2, January 2010). Both TerraSAR-X and TanDEM-X were designed for a five-year life span. As of 2019, both satellites were still on orbit (more than 10 years in operation) and were expected to continue their mission lives for at least another four years. The radar specifications for TerraSAR-X include[24]:

Operational life	5 years, for both satellites (TerraSAR-X and TanDEM-X), an extended lifetime of that least another 5 years (beyond 2018) is expected by the operator DLR (Status: April 2014)
Orbit	Sun-synchronous repeat orbit
Repeat period	11 days
Equatorial crossing time (GMT)	18:00 h ascending pass (±0.25 h) 06:00 h descending pass (±0.25 h)
Inclination	97.44°
Altitude at the equator	514 km (319.8 miles)
Antenna type	Active phases array antenna, electronically separable
Antenna size	4.78 m × 0.7 m (15.7 feet × 2.3 feet)
Centre frequency	9.65 GHz (X-band)
Chirp bandwidth	150 MHz/300 MHz
Nominal acquisition direction	Right side
Polarization	Single, dual—depending on imaging mode quadruple is available as advanced polarization mode for dedicated acquisition campaigns

1.3.3 ImSAR

ImSAR,[25] out of Utah in the United States, specializes in lightweight SAR for medium-to-large-sized aircraft. Operating in the Ku band, the ImSAR NanoSAR Pod (NSP) provides a multitude of coherent and incoherent SAR products. These include standard NITF images, compressed image formats such as JPG, PNG, and BMP, coherent and incoherent change detection, maritime surveillance, and moving target indication (MTI) tracks. The SAR is capable of spotlight and stripmap modes, as well as a wide-area search mode, which can monitor an area using a gridded imaging technique. The SAR can be operated in either 0.3-meter or 1.0-meter resolutions, weighs approximately 24 pounds, and uses about 275 watts of power; the NSP-7 can be operated at up to 23,000 feet above ground level (AGL). Figure 1.6 shows an excellent image of the Las Vegas, NV area, taken with the ICEYE-X2 satellite on January 6, 2019. The airport and the strip are clearly visible in the top center of the image.

---

[24] Source: `https://www.satimagingcorp.com/satellite-sensors/terrasar-x-radar-satellite/`

[25] `https://www.imsar.com`

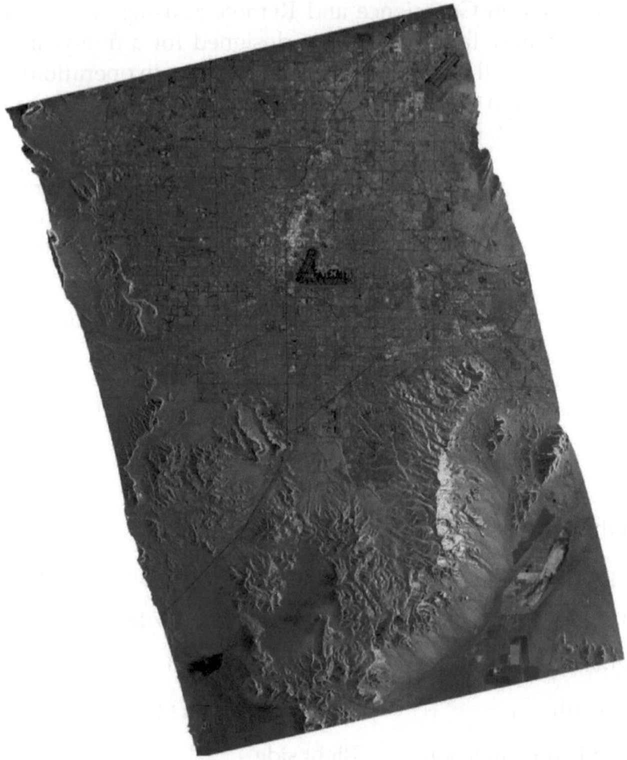

Figure 1.6 ICEYE-X2 image of Las Vegas, NV, taken on January 6, 2019. This is a single-channel, HH polarization image. (*Image credit: ICEYE. Used with permission.*)

1.3.4 ICEYE

ICEYE[26] is a Polish and Finnish company focused on small satellite design, launch, and operation. It is the first organization in the world to launch a space-based radar under 100 kg and the first Finnish commercial satellite. Its overarching goal is to produce a constellation of small sats that could image the entire planet once a day, everyday. The first payload, called ICEYE-X1, was launched on January 12, 2018. The X1 payload has a radar bandwidth of 60 MHz, giving it a nominal ground resolution of 10 meters. The polarization for the system is a single channel VV (where transmit polarization is vertical, and the receive polarization is also vertical). X1 was capable only of stripmap imaging and was considered a pathfinder satellite.

ICEYE-X2 was launched on December 3, 2018. While slightly larger in size and weight than X1, it delivers a good deal more capability, including the ability to image in stripmap, spotlight, and scan, three different resolution modes (3-meter in stripmap, 20-meter in scan, and 1-meter spotlight). The radar bandwidth is 300 MHz, which supports the spotlight resolution of 1-meter.

[26] https://www.iceye.com

FIGURE 1.7 One of the first Capella images, released on October 6, 2020. The image is of Palm Jumeirah in Dubai and exhibits significantly high dynamic contrast, particularly in the maritime areas around the Palm. (*Image credit: Capella Space. Used with permission.*)

1.3.5 Capella Space Systems

The first U.S. commercial space-based SAR venture was created by Capella Space Systems, which was founded in 2016. The company launched its first smallsat SAR, part of its Sequoia constellation in August 2020. The Capella Sequoia constellation, which will consist of 36 satellites, is made up of an X-band, single-frequency radar.[27] The payload can image in spotlight, sliding spotlight (sometimes called ScanSAR), and stripmap modes. The spotlight mode covers an area of 5 km × 5 km, the stripmap mode covers a strip of 5 km × 20 km, and the sliding spotlight mode (called "Site" in the Capella Products Guide[28]) covers an area of 5 km × 10 km.

The radar has a center frequency of around 9.5 GHz and transmits with an average power level of 600 watts. It provides single transmit and receive polarizations in horizontal (HH). The satellites can maneuver quickly with a rotation of 1.5° per second, allowing the sensor to image on either the left or right sides of the orbit ground track. The company delivers either single-look complex (SLC) data or geocoded (GEO) data. The resolution for the ground-projected GEO product for spotlight imaging is 0.5 meters, while the sliding spotlight delivers a 1.0-meter resolution, and stripmap imaging gives a 1.7-meter resolution. The system is capable of imaging with look angles of 25° to 40° degrees. This is the angle that is measured from the normal to the ground plane and the slant plane. This is the complement to grazing angle, which is created by the slant and ground planes. The geometry of SAR is discussed in the next section.

Figure 1.7 shows one of the first Capella images of Palm Jumeirah in Dubai. Taken in stripmap mode, the image has a resolution of about 2.0 meters and shows exquisite detail, particularly in the wave features in the Arabian Gulf around the Palm.

[27] D. Werner, "Capella Space unveils first radar satellite images," *SpaceNews*, October 7, 2020.

[28] Capella Space, *SAR Imagery Products Guide*, Version 2.16, September 11, 2020.

1.4 Timing and Geometry

Arguably, the most important aspect of SAR is geometry and timing. Without careful accounting of the timing, the coherency of the waves that are transmitted and received cannot be maintained. Likewise, precise knowledge of the geometry is needed for *focusing* the SAR image. Throughout this book, *focusing* will refer to a series of algorithms that will be used to take the transmitted waveform and extract a radar reflectivity map of the entire scene. This series of algorithmic steps is typically called *image formation* or *image formation processing*. How precise the geometry and timing need to be depends highly on the resolution of the system, the imaging mode, the imaging geometry, and the desired quality of the final image. Another term that is frequently used for focusing is called *compression*.

1.4.1 Imaging Modes

Generally, there are three different imaging modes that are employed in SAR. These are illustrated in Figure 1.8. The first, called *stripmap*, fixes the antenna to the side of the aircraft or spacecraft and drags the beam across the ground. This limits the maximum azimuth resolution to the beamwidth associated with the antenna. The first SAR systems used stripmap mode.

The next mode is called *spotlight*. As the name implies, the beam is fixed to a certain spot on the ground and the antenna is slewed, either mechanically or electronically, so as to keep the beam in that spot. This increases the maximum azimuth resolution, limiting it only by the total change in look angle (called squint angle). Theoretically, if it could view all of the angles around a scene, a spotlight system could achieve an azimuth resolution on the order of a fraction of the radar wavelength. The disadvantage is that the area a spotlight system can cover is limited by the spot on the ground, which is determined by the antenna beamwidth.

Finally, the last mode is called *scan* and is a hybrid between stripmap and spotlight modes. In scan mode, the beam is both dragged and slewed to perform at higher resolutions than can be achieved with stripmap, but cover more area than what can be achieved with spotlight mode.

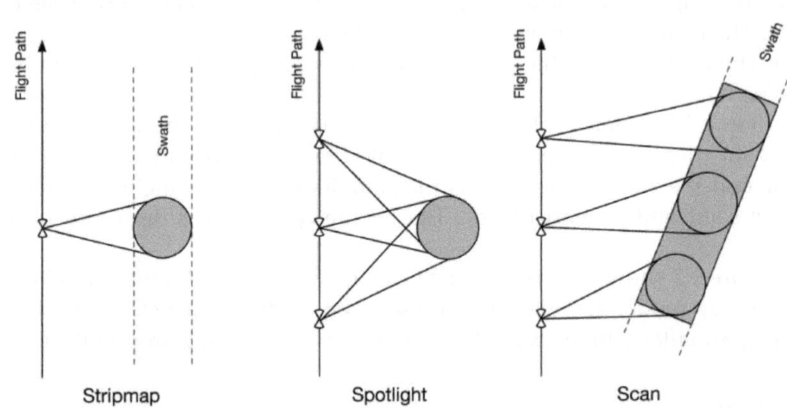

FIGURE 1.8 The three imaging modes of synthetic aperture radar: stripmap, spotlight, and scan.

1.4.2 Geometry

Stripmap imaging mode will be used first to describe the geometric model for SAR. Figure 1.9 shows the overhead geometric model for stripmap, where the independent axis represents range or *fast time*, and the dependent axis represents azimuth or *slow time*. The closest range to the radar in the beam pattern is called the *near edge* and the furthest range to the radar is called the *far edge*. These edges are defined by the illumination pattern of the antenna onto the ground and make up the extents of the swath or scene. The center of the swath is called the *ground reference line* (GRL). The center of the swath can also be arbitrary defined as the *ground reference point* (GRP) and is mathematically defined here as $(0, 0, 0)$. The ground reference line intersects the ground reference point. The center of the antenna, which corresponds to the aircraft or spacecraft's location, is defined as the *antenna reference point* (ARP) and is defined mathematically as (x_a, y_a, z_a).

If the antenna is not looking perpendicular to the flight track and is, instead, looking either forward or backward, this is called "squinting." The angle between the range vector to the center of the beam and the vector perpendicular to the flight track is called

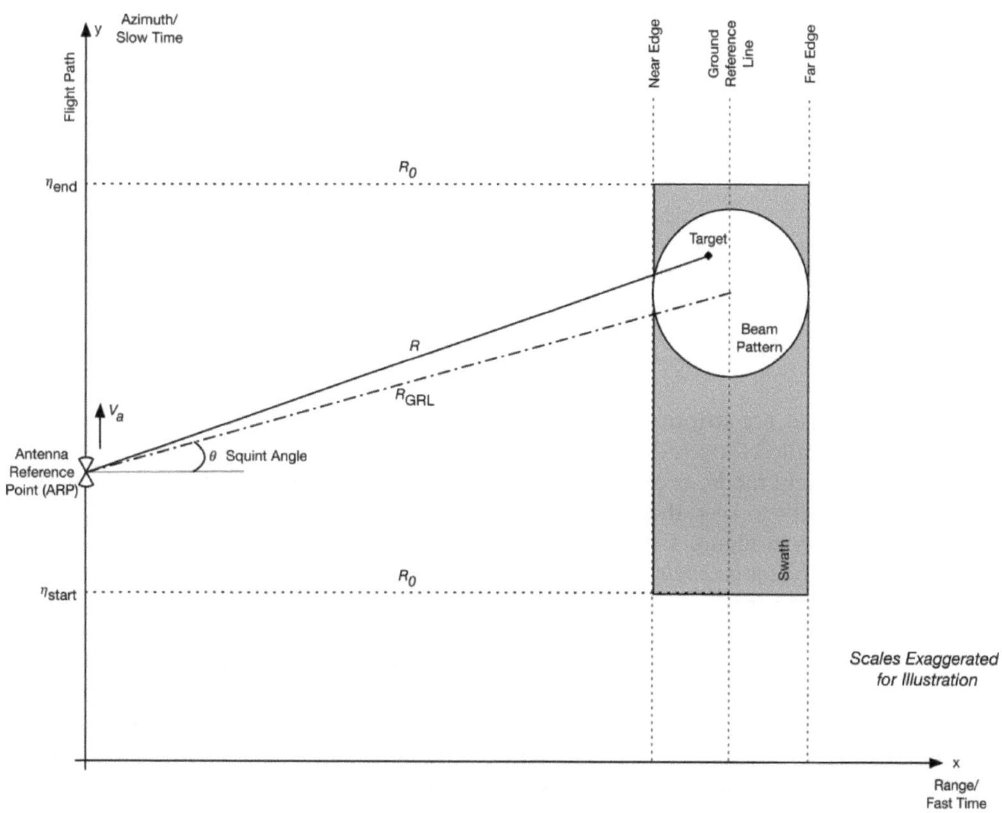

FIGURE 1.9 Stripmap mode with relevant ranges and angles highlighted.

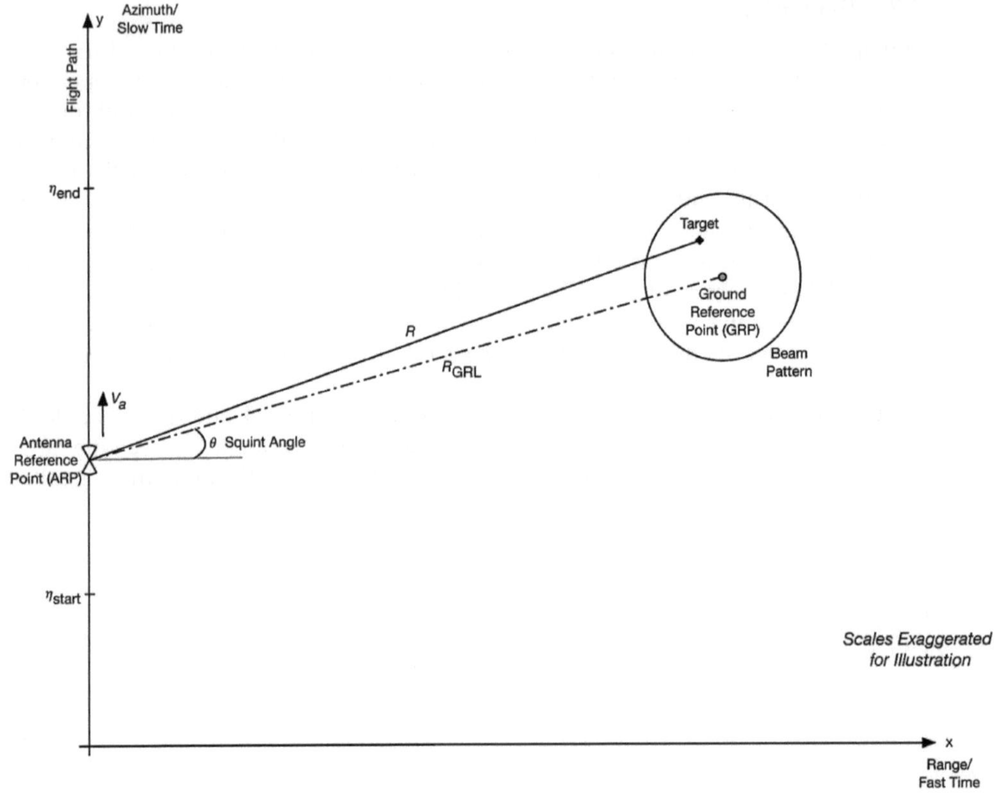

Figure 1.10 Spotlight mode with relevant ranges and angles highlighted.

the *squint angle*. For stripmap, this angle is fixed. The vector perpendicular to the flight track is also called the *zero-Doppler vector*.

For spotlight mode, as mentioned earlier, the antenna is slewed, either electronically or mechanically, to keep the beam centered on the GRP. The geometric model for spotlight is shown in Figure 1.10. Again, the angle created from the center of the beam to broadside is the *squint angle*. Unlike stripmap, the squint angle in spotlight changes over the course of the entire dwell.

Figure 1.11 illustrates the imaging geometry as seen from a side view. The x-axis shows range along the ground and the z-axis is altitude or elevation above the ground. The scene is the patch of ground between the near edge and the far edge, as shown in the figure. The GRP is naturally in the middle. The angle created between the range vector at the GRP and the ground plane is called the *grazing angle* and is denoted by the variable ψ. The grazing angle is typically measured at the GRP for spotlight or the GRL for stripmap. The angle created between the near edge range vector and the far edge range vector is called the *antenna beamwidth*. The beamwidth is also sometimes described by width of the beam on the ground in either range or azimuth.

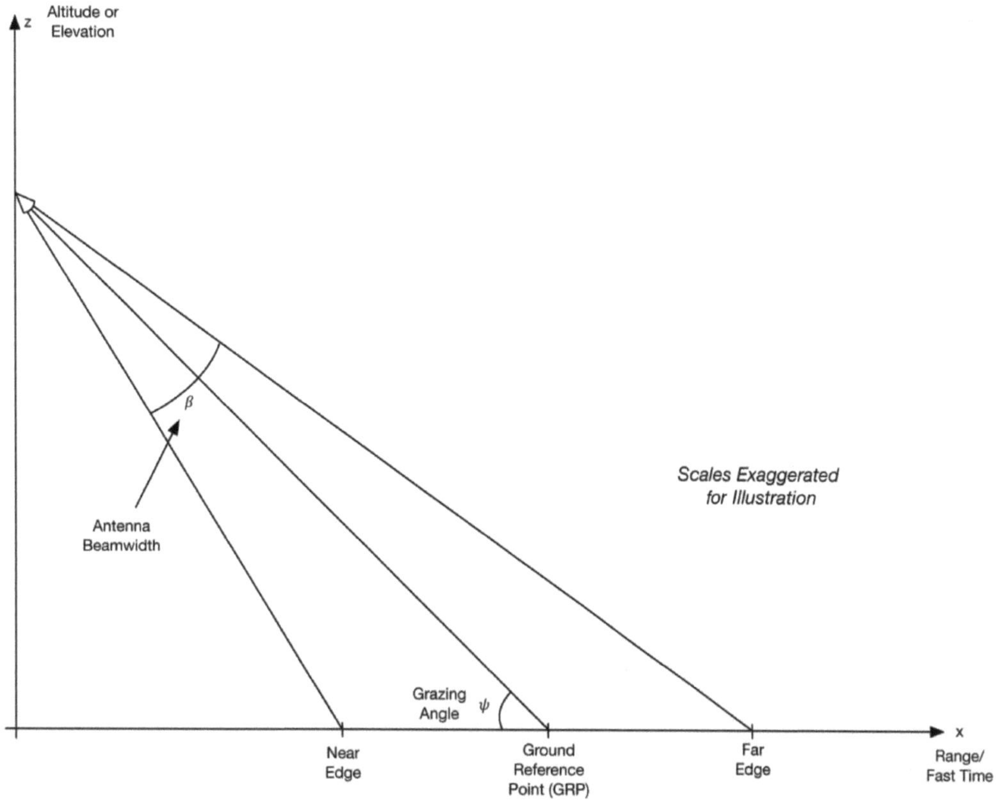

FIGURE 1.11 Side view of imaging geometry.

Example 1.1 Consider an aircraft flying at 20,000 feet AGL. The sensor is known to have a fixed grazing angle of 55° (measured from the ground up) at the ground reference point. The distance from the ground track (i.e., the flight track projected onto the ground plane) to the center point is, using the trigonometric functions of a right triangle,

$$x = \frac{z}{\tan(\psi)} = \frac{20{,}000}{\tan(55°)} = 14{,}004.15 \text{ feet} = 4.27 \text{ km} \tag{1.1}$$

HINT: It may be useful to examine Figure 1.11 for help in visualizing the geometry.

1.4.3 Timing

There are two different timing dimensions: *fast time* and *slow time*. These dimensions colloquially describe the timing that happens along a pulse and along a collection of pulses, respectively. That is, fast time is the period of time that is needed to sample the returned signal, which is typically the period of time that it takes the round-trip signal to travel

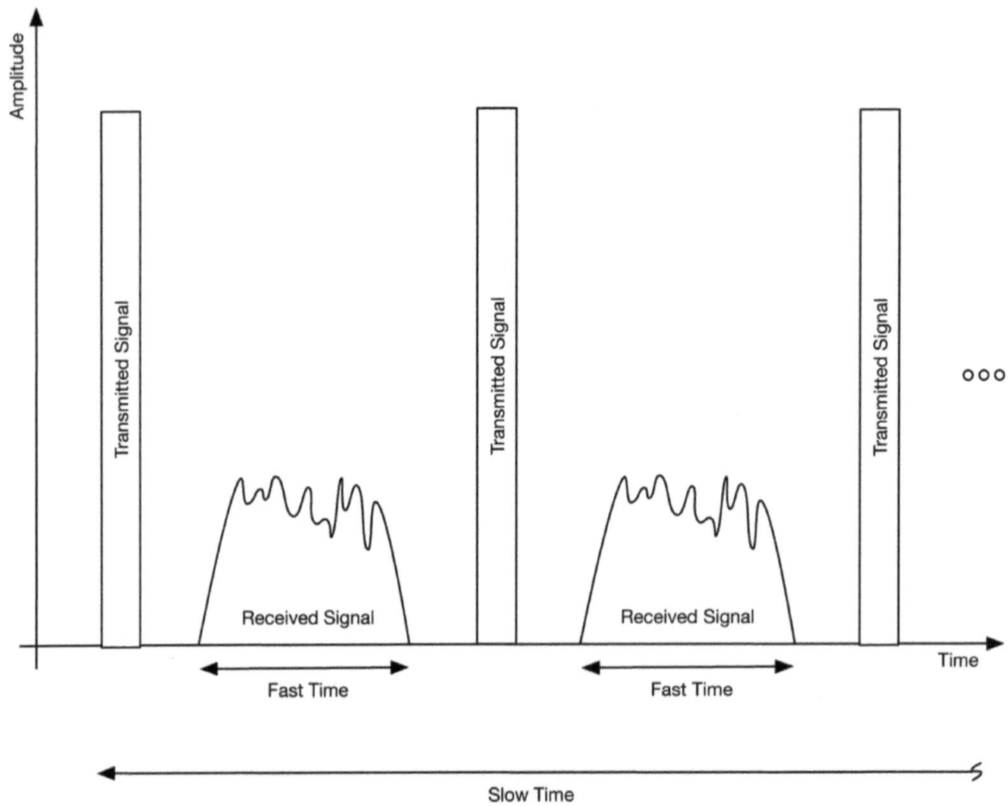

FIGURE 1.12 A notional depiction of fast and slow times.

across the scene and return to the radar. In a pulsed radar system, this occurs between transmitted pulses, as shown in Figure 1.12. Conversely, slow time is the collection of returns after each transmitted pulse. Notationally, fast time is represented here by the variable t and slow time is represented by the variable η.

Figure 1.13 shows the grid of stacked fast time samples. Each fast time vector is stacked consecutively in slow time. The duration of fast time samples is on the order of microseconds or milliseconds, depending on the round-trip delay times. The duration of slow time is on the order of seconds or tenths of a second, depending on the amount of time needed to form the synthetic aperture (more on this in the next section).

Consider the geometry shown in Figure 1.14. This geometry illustrates a sensor traveling at a constant velocity v_a along a straight and level flight path. The sensor is shown at two different points in slow time. The angle that is made between any given range and the ground is called the grazing angle. This angle is typically reported from the scene center and the center of the dwell for spotlight SAR. It is constant over the dwell for stripmap SAR. The plane that is created between the flight track and the ground for a set of range is called the *slant plane*. The plane that is created from the intersection of the nadir line, the zero-Doppler vector, and the slant plane is called the *zero-Doppler plane*.

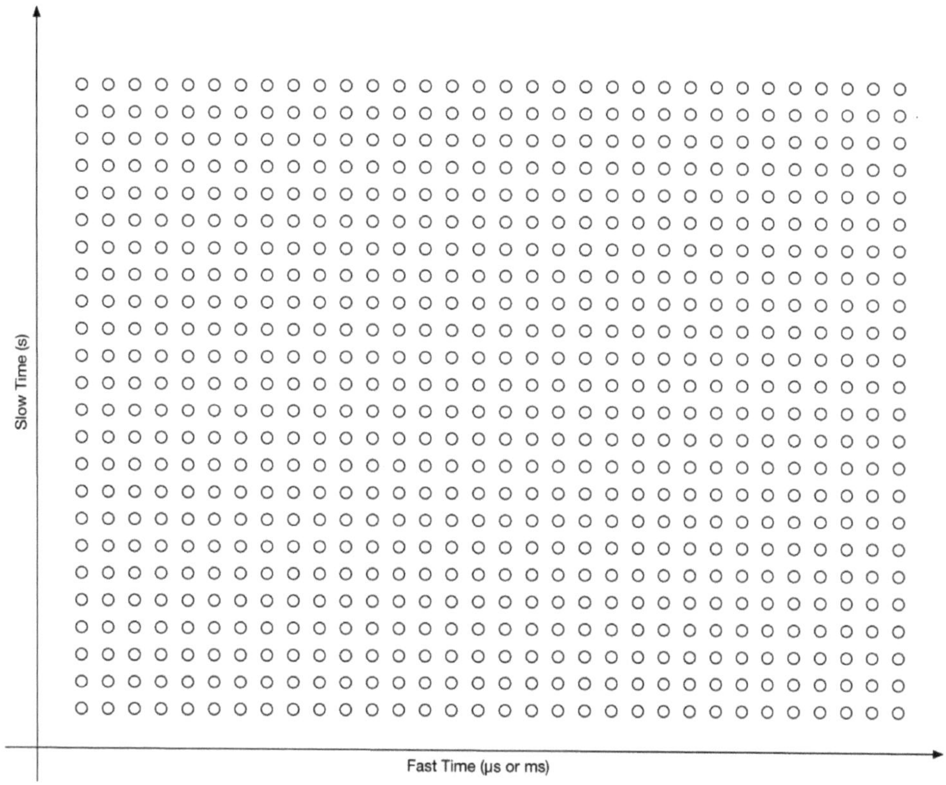

FIGURE 1.13 The grid of stacked fast time samples along slow time.

Euclidean distance is typically used to calculate the range between the sensor and any given point in the scene. Recall that Euclidean distance, in three dimensions, is defined as

$$R(x,y,z) = \sqrt{(x_1 - x_2)^2 + (y_1 - y_2)^2 + (z_1 - z_2)^2} \tag{1.2}$$

where (x_1, y_1, z_1) and (x_2, y_2, z_2) are the reference points between which distance $R(x, y, z)$ is measured.

Example 1.2 Figure 1.15 shows an example of the range between a hypothetical target at $(10, 10, 0)$ meters from the scene center and the range between the sensor and the scene center (also known as the ground reference point). The aircraft is traveling at a constant 80 meters per second, while flying straight and level at an altitude of 3000 meters above ground level (AGL).

The range between the sensor and the GRP is crucial to the formation of the SAR image, as all of the ranges in the scene are typically referenced to it. The stars in the curves indicate the *closest approach* between either the sensor and the target or the sensor and the scene center. The term "closest approach" refers to the closest distance between the sensor and any arbitrary target in the scene, which includes the

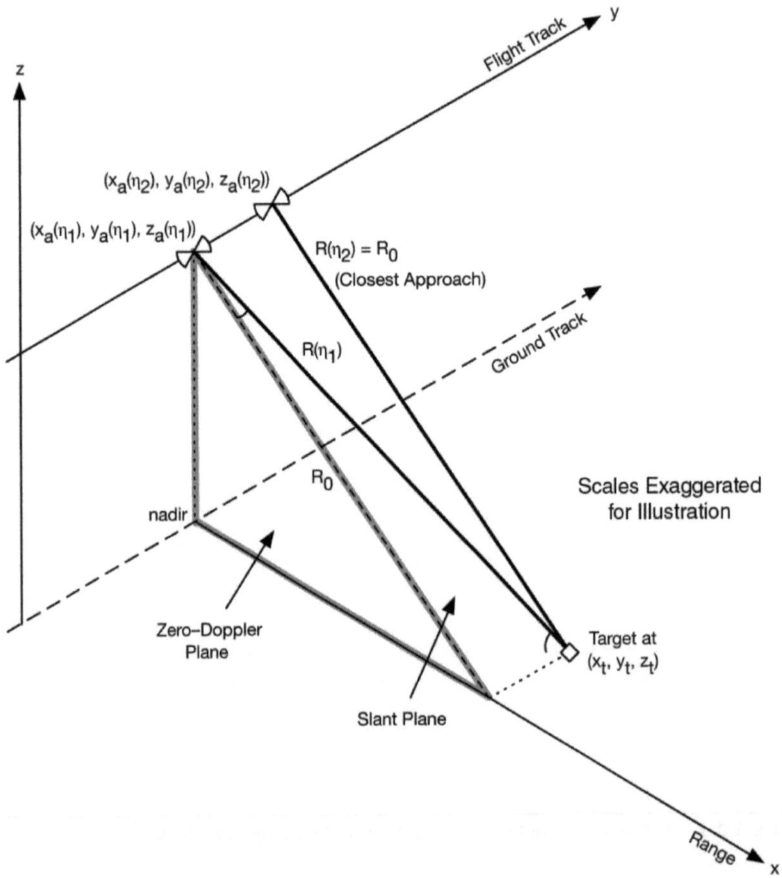

FIGURE 1.14 A 3D view of the sensor, traveling at some constant velocity, at two different points in slow time. The ranges are in reference to a hypothetical target in the scene, as well as to the zero-Doppler vector.

GRP. Knowledge of the closest approach may be crucial for some image formation algorithms.

Example 1.3 Refer to Figure 1.16 to consider a SAR system that has a known 3-dB azimuth beamwidth of 12.5°. The distance from the sensor to the ground reference point is 5 kilometers. The width of the beam at the ground reference point is

$$y_w = 2\left(\tan\left(12.5^\circ\right) 5000\right) = 2216.9 \text{ m} = 2.21 \text{ km} \tag{1.3}$$

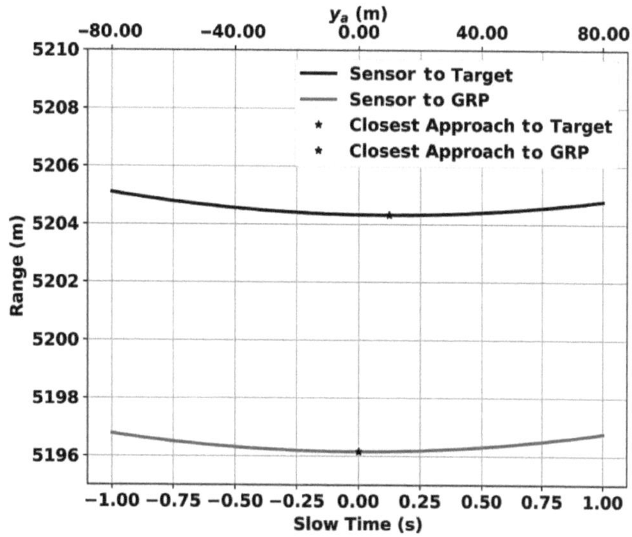

FIGURE 1.15 Range to a target at (10, 10, 0) meters from scene center for a sensor onboard a hypothetical aircraft going 80 meters per second at an altitude of 3000 meters.

1.5 The Synthetic Aperture

A natural question that may be asked is, "What is the synthetic aperture and why is it important?" To answer this, it is necessary to discuss *real aperture radar* (RAR).

As shown in Figure 1.3, targets along the same iso-range line cannot be distinguished from one another. However, consider the possibility of shrinking the width of the antenna beam pattern on the ground so that it only covers the width of the desired resolution. Then the beam could be scanned across the ground and each pulse would contain the information of each individual target, provided the pulses were spaced so that there were no gaps and not a significant amount of overlap. This is called *real aperture radar.*

Figure 1.17 shows the concept of RAR. The cross-range beamwidth is decreased so that the width proportional with the desired cross-range resolution. The scene width is designated as D, while the azimuth beamwidth is designated as W_a. The azimuth beamwidth is defined mathematically as

$$W_a = R\frac{\lambda}{L_a} \tag{1.4}$$

where R is the range from the sensor to the furthest edge of the scene, λ is the wavelength of the sensor, and L_a is the physical width of the antenna in the azimuth direction. The wavelength of the system is related to the center frequency of the system by the relationship

$$f_0 = \frac{c}{\lambda} \tag{1.5}$$

where f_0 is the center frequency of the radar and c is the speed of light.

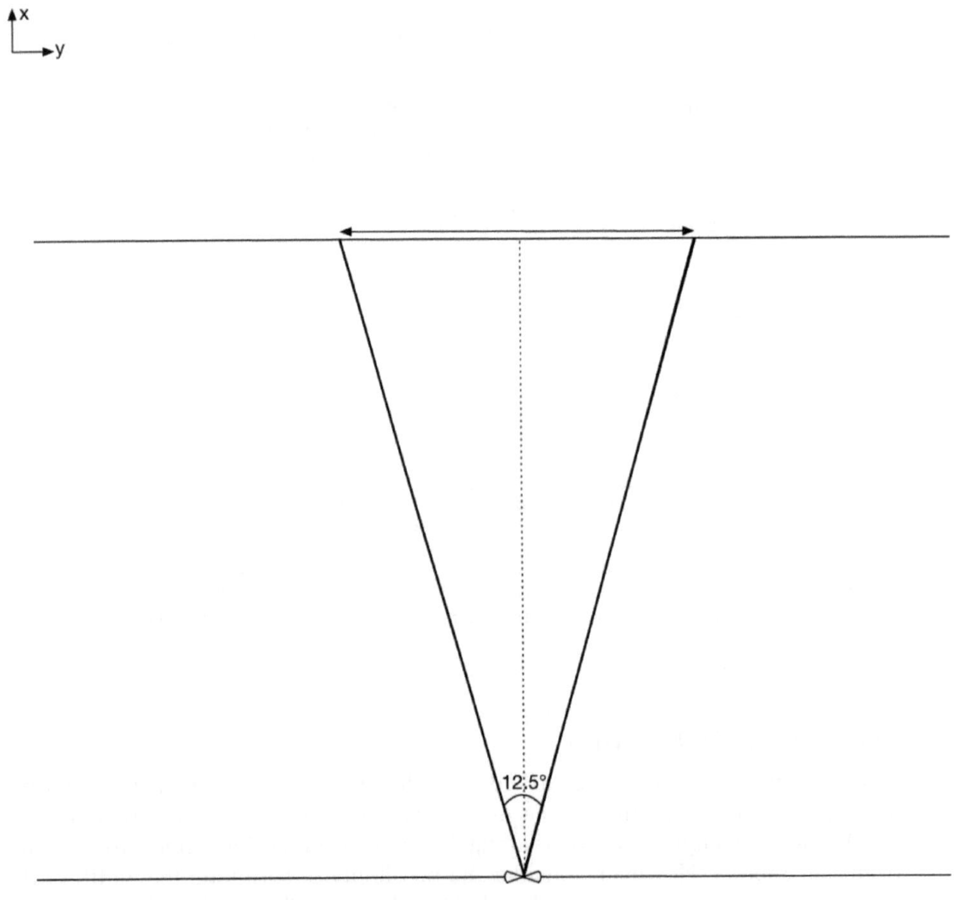

Figure 1.16 Geometry for considering antenna beamwidth in azimuth direction.

Example 1.4 Consider an X-band radar system with a center frequency of $f_0 = 10$ GHz with a nominal range of 50 kilometers. The wavelength of such a system is

$$\lambda = \frac{c}{f_0} = \frac{299{,}792{,}458 \text{ m/s}}{10 \times 10^9 \text{ Hz}} \approx 0.03 \text{ m} = 3 \text{ cm} \tag{1.6}$$

If an azimuth resolution of 0.5 meters is desired, then

$$L_a = \frac{R\lambda}{W_a} = \frac{50 \times 10^3 \text{ m} \cdot 0.03 \text{ m}}{0.5 \text{ m}} = 3000 \text{ m} \tag{1.7}$$

An antenna width of 3 kilometers is needed to achieve the azimuth resolution of a half meter! Practically speaking, the antenna would need to be length of nearly 33 football fields. This is clearly impractical. Generally speaking, high-resolution imaging radar is impractical with RAR, where high resolution in this context means resolutions finer than a kilometer scale.

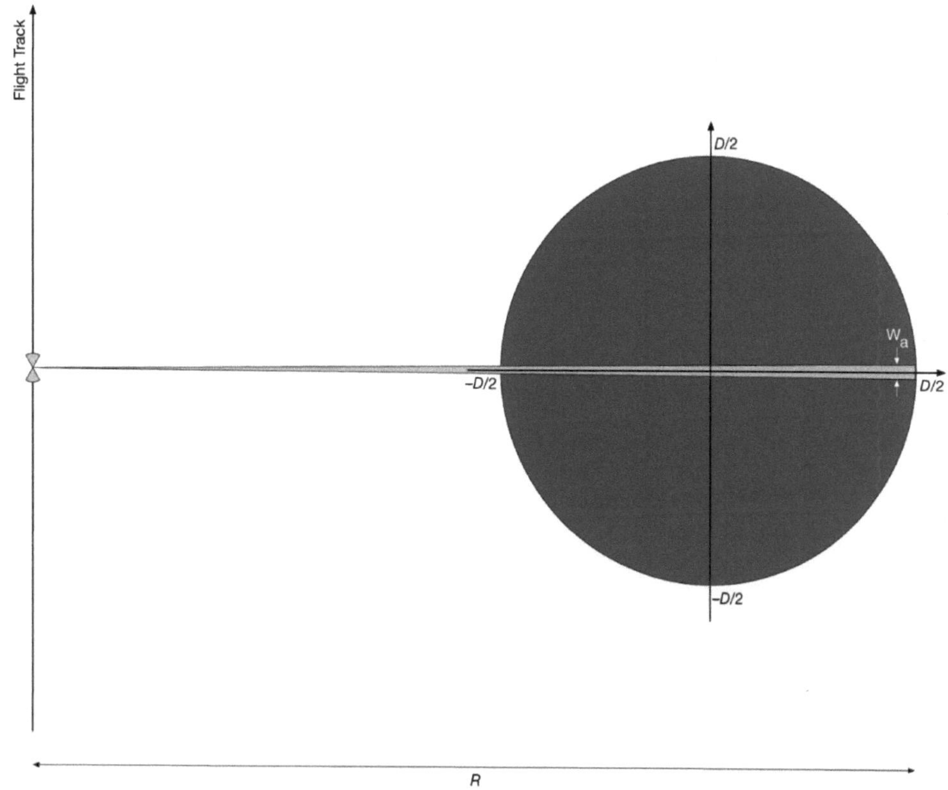

FIGURE 1.17 Geometry of real aperture radar (RAR).

For fine resolution imaging, another technique is needed to generate the azimuth resolutions desired. This is where the *synthetic aperture* comes in. The synthetic aperture is used to create an artificial antenna that can resolve objects at the high resolution required. Chapters 2 and 3 will discuss how ranging is accomplished and how the synthetic aperture is formed.

CHAPTER 2
Ranging

The primary function of a radar is to determine distance between the instrument and an object or set of objects. This chapter will introduce the mathematical foundation for ranging with a synthetic aperture radar. The reasoning behind using a linear frequency-modulated (LFM) chirp pulse will be given. Two techniques for demodulating the received radar signal will be given, as well as advantages and disadvantages for both.

2.1 Range via Echoing

The technique of using echoing for determining the range of objects is used in a number of technologies, including radar (radio waves), LIDAR (light waves), and sonar (sound waves). This is done by transmitting an energy wave, which can either be a pulsed or continuous wave, and listening for its echo.

Figure 2.1 is a simple illustration of how ranging works. The diagram shows a radar sensor at some arbitrary altitude. The sensor transmits a pulse and that pulse travels at very nearly the speed of light toward the ground; since the wave is traveling through a medium and not a vacuum, the velocity is slightly less than the speed of light.[1] The time it takes between transmission and reception is the round-trip time delay, and it is represented by τ. The distance between the sensor and the object is R, which is typically represented by Euclidean distance.

Range is related to the round-trip time delay by

$$R = \frac{c\tau}{2} \tag{2.1}$$

where R is range, c is the speed of light, and τ is the time delay from when the waveform leaves the sensor to when it hits the object, and bounces back to the sensor. Thus, if the distance from an antenna to an object is 1500 meters, then the time delay between the antenna and the object is

$$\tau = \frac{2R}{c} = \frac{2(1500\text{ m})}{3 \times 10^8\text{ m/s}} = 10\ \mu\text{s}. \tag{2.2}$$

[1] It should be noted that the wavefront is actually curved as it travels from the sensor to the ground.

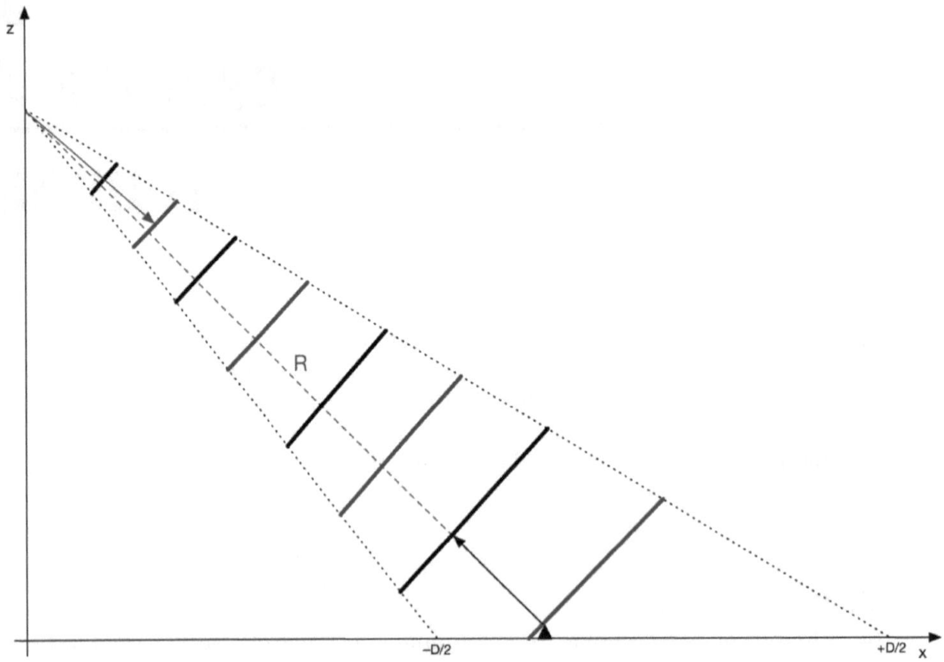

FIGURE 2.1 Simple diagram illustrating ranging.

The most common pulse that is transmitted for synthetic aperture radar is called the linear frequency-modulated (FM) signal. This is sometimes called a "chirp." The chirp will be described in this chapter along with two different techniques to compress the pulse. *Pulse compression* is a digital communication technique. It is a type of *spread spectrum* method, which minimizes power requirements while maximizing the signal-to-noise ratio (SNR). Pulse compression is required for high-resolution synthetic aperture radar; simple single-sinusoid pulses cannot achieve the necessary bandwidth while operating at a realistic power level.

2.2 The Linear Frequency-Modulated Chirp

The linear frequency-modulated (LFM) waveform, or "chirp," is common in synthetic aperture radar. It is typically a pulsed sinusoid that starts at one frequency and increases ("up chirp") or decreases ("down chirp") in frequency across the width of the pulse. The rate of change in frequency is called the *chirp rate*, and it is represented here by K and is typically in units of hertz per second.

The LFM chirp is represented by

$$s(t) = A_t \operatorname{rect}\left(\frac{t}{T}\right) \cos\left(2\pi f_0 t + \pi K t^2\right) \tag{2.3}$$

where t is fast time in seconds, A is the transmitted amplitude of the LFM, T is pulse width in seconds, f_0 is the center frequency of the radar in hertz, K is the chirp rate in hertz per second, and rect() is the rectangle function, which is defined as

$$\mathrm{rect}(t) = \begin{cases} 0 & \mathit{for} \quad |t| > \frac{1}{2} \\ \frac{1}{2} & \mathit{for} \quad |t| = \frac{1}{2} \\ 1 & \mathit{for} \quad |t| < \frac{1}{2} \end{cases} \tag{2.4}$$

In essence, the rect() function defines the pulse and the cos() defines the shape of the waveform inside the pulse, in this case, an LFM chirp with a center frequency of f_0 and a chirp rate of K. When K is positive, this corresponds to the up chirp; conversely, the down chirp occurs when K is negative.

Figure 2.2 shows the time series form of the chirp along with the phase of the chirp and the time versus frequency plot. This is for a chirp with a carrier frequency at DC (0 Hz). As the carrier frequency increases, the number of cycles between the start and end of the pulse increases. DC is chosen here for illustration purposes. A typical radar system will use a carrier wave in the high megahertz to low gigahertz range. For example, the TerraSAR-X radar system has an advertised center frequency of 9.65 GHz, which is in the X-band (thus, the "X" at the end of the TerraSAR-X name). The width of the pulse in frequency space is called the *bandwidth*. The radar bandwidth is a crucial parameter, as it controls the resolution of the radar in the range direction. Range resolution is inversely proportional to bandwidth, so the greater the bandwidth, the finer the range resolution. Put simply, better resolution requires greater bandwidth.

Note that the phase takes the form of a quadratic. Recall from Chapter 1 that the range between any given object in the scene and the sensor is also quadratic in slow time. So, any arbitrary object in the scene will produce a quadratic in phase in both fast time and slow time. The form of these quadratics will be used to detect any given scatterer using a *matched filter*. The matched filtering techniques used to detect these quadratics will vary; several of the techniques will be outlined in this book.

Once a pulse is transmitted, the radar listens for the return pulses. This is often performed by the same antenna that transmitted the pulse. A switch is used to direct the flow of signal traffic between the transmitter and the receiver. Once received, the chirp is now of the form

$$s_r(t) = A_r \,\mathrm{rect}\left(\frac{t-\tau}{T}\right) \cos\left(2\pi f_0(t-\tau) + \pi K(t-\tau)^2\right) \tag{2.5}$$

where A_r is the attenuated signal from an arbitrary object's reflection and τ is the round-trip time delay between the sensor and the object. Radar has a range loss inversely proportional to range to the 4th power ($\frac{1}{R^4}$). This derives from the radar range equation,

$$P_r = \frac{P_t G^2 \lambda^2 \sigma}{(4\pi)^3 R^4} \tag{2.6}$$

where P_r is the received power (in watts), P_t is the transmitted power (in watts), G is the antenna gain (in decibels; G^2 assumes that the same antenna is used for both transmit and receive or that if two different antennas are used, they have the same gain), λ is the wavelength of the radar (in meters), and R is the range from any given object to the sensor

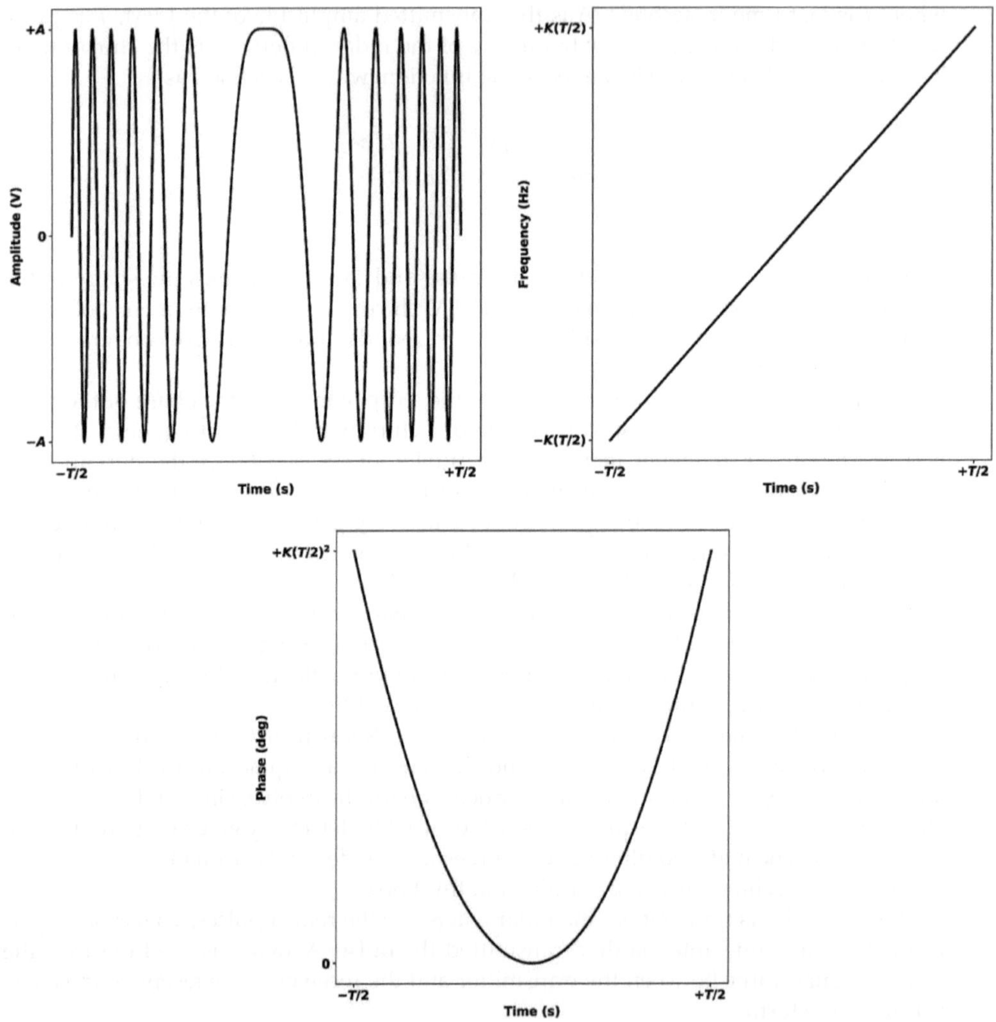

FIGURE 2.2 A linear frequency-modulated (LFM) chirp pulse (top left), its change in frequency over time (top right), and its change in phase over time (bottom).

(in meters).[2] Note that once SAR image formation processing has been performed, the integration of pulses over slow time relaxes the range requirement from R^4 to R^3, which helps with power requirements for the system.[3]

The form of Equation (2.5) indicates that during operation, the radar receives an attenuated, time-delayed LFM chirp from the reflection of some given object in the scene.

[2] M. Skolnik, *Radar Handbook*, Third Edition. McGraw-Hill Education, New York, NY, 2008.

[3] W. G. Carrara, R. S. Goodman, and R. M. Majewski, *Spotlight Synthetic Aperture Radar: Signal Processing Algorithms*. Artech House, Norwood, MA, 1995.

In reality, the received signal is a mixture of reflections from all of the objects in the scene. This is a superposition of LFM chirps, defined as

$$s_r(t) = \int_{x=-D/2}^{D/2} B(x)\,\mathrm{rect}\left(\frac{t-\tau(x)}{T}\right)\cos(2\pi f_0(t-\tau(x)) + \pi K(t-\tau(x))^2)\,dx \qquad (2.7)$$

where D is the width of the scene in the range direction. In frequency space, this signal is centered at the carrier frequency of the radar. The next section will outline two different techniques for demodulating the signal and detect the demodulated chirps at *baseband*.

2.3 Detecting the Received LFM Chirps

For this section and for simplicity, let's consider only one reflected LFM chirp from an arbitrary target in the scene. This takes the form of Equation (2.5). There are two specific techniques for demodulating this signal: *quadrature demodulation* and *dechirping*. Quadrature demodulation is a common technique used in communications to convert a signal back to baseband.[4] Dechirping (also called stretch processing[5]) involves demodulating with the original chirp, time delayed to the scene center, also called the *ground reference point* (GRP). As will be shown, quadrature demodulation leaves a copy of the time-delayed chirp at baseband and requires matched filtering to detect that chirp. For dechirping, the matched filtering is built-in; the corresponding output is a sinusoid whose frequency is correlated with the distance of the object to the sensor.

2.3.1 Quadrature Demodulation

Quadrature demodulation is a technique that mixes a received signal with a sinusoid at the carrier frequency. The mixed signal is low-pass filtered to remove the copy of the signal at twice the carrier frequency, leaving only the baseband signal. The same mixture is performed using a sinusoid 90° out of phase. The signals are digitally sampled and combined as a complex number to create the quadrature. This is done to remove the innate ambiguity that comes with sinusoidal signals and tracking their phase: it is not possible to know whether the signal is on the uptick or the downtick of the sinusoid. This ambiguity manifests itself in Fourier analysis as positive and negative frequencies. Quadrature processing creates a phasor that removes this ambiguity, leaving one impulse per sinusoid at a given frequency.

Figure 2.3 illustrates the basic principle of phasors. A complex number is a phasor that provides two pieces of information that can be combined to get the magnitude and phase of a vector. For quadrature demodulation, the two pieces of information, the unfortunately named "real" and "imaginary" components, are generated after the received signal is mixed with a cosine and sine at the carrier frequency and then a low-pass filter is

[4] R. G. Gallager, *Principles of Digital Communication*. Cambridge University Press, Cambridge, 2008.

[5] C. V. Jakowatz, D. E. Wahl, P. H. Eichel, and D. C. Ghiglia, *Spotlight-Mode Synthetic Aperture Radar: A Signal Processing Approach*. Elsevier, 1996.

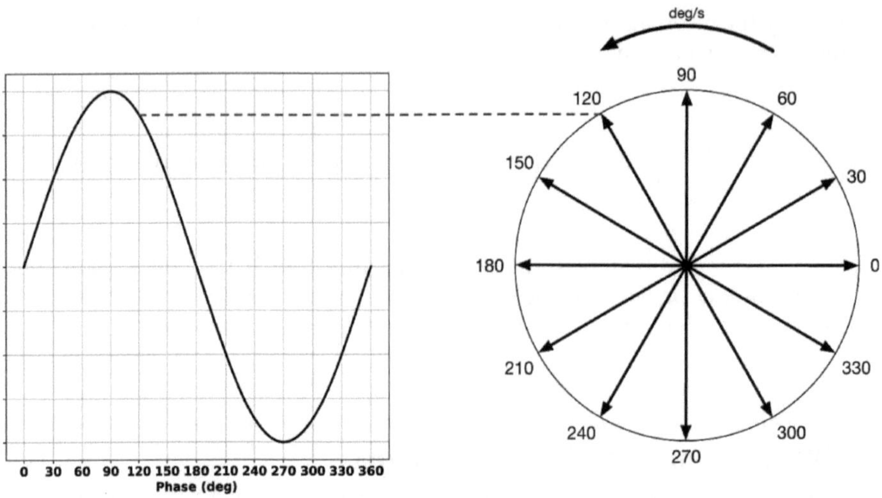

FIGURE 2.3 A graphical description of phasors. A complex number essentially is a phasor that provides both magnitude and phase information.

used to eliminate all frequencies except those at baseband. The magnitude is calculated by

$$m = \sqrt{I^2 + Q^2} \tag{2.8}$$

and the phase is calculated by

$$\phi = \tan^{-1}\left(\frac{Q}{I}\right) \tag{2.9}$$

Figure 2.4 shows the flow of quadrature demodulation processing. This is all performed on analog signals until the signal is digitized. The received signal is mixed with a single cosine at the carrier frequency, as well as sine at the same frequency. These are called the *reference signals*. The phases of the reference signals are tightly controlled; if they are not well-controlled, the output of the matched filter will not be optimal. This nonoptimality will appear as a *phase error* and, if not corrected, may affect image quality. In this case, the phase error corrupts the estimated range between the sensor and the object. Depending on the severity of the phase error and its consistency, the phases from pulse-to-pulse may be affected.

Recall that a time-delayed LFM chirp is

$$s_r(t) = \cos(2\pi f_0(t - \tau) + \pi K(t - \tau)^2) \tag{2.10}$$

We'll temporarily ignore the attenuation term and the rect() for convenience. Note that this can be written as

$$s_r(t) = \cos(2\pi f_0 t - 2\pi f_0 \tau + \pi K(t - \tau)^2) \tag{2.11}$$

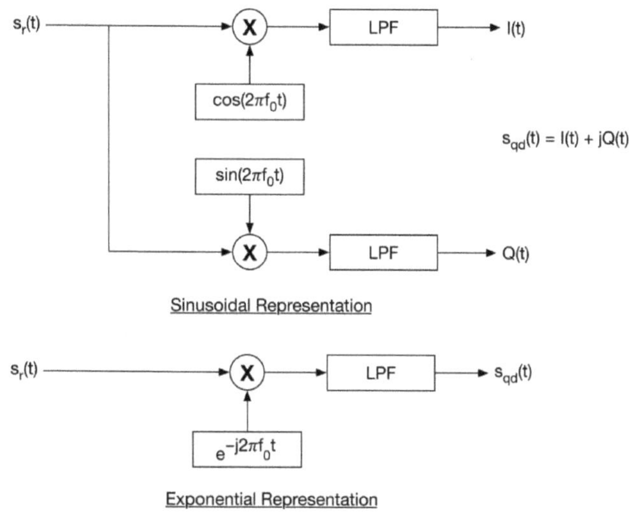

FIGURE 2.4 Two flowcharts illustrating quadrature demodulation. The top flow shows demodulation using sinusoidal notation; the bottom flow shows demodulation using exponential notation. The flowcharts are equivalent.

Now, let $s_r(t) = \cos(2\pi f_0 t + \Phi(t))$ and $h(t) = \cos(2\pi f_0 t)$. Here $h(t)$ serves as our reference signal and $\Phi(t) = -2\pi f_0 \tau + \pi K(t - \tau)^2$. Figure 2.4 shows $s_r(t)$ and $h(t)$ being mixed, which is mathematically equivalent to multiplying the signals together,

$$s_r(t) \cdot h(t) = \cos(2\pi f_0 t + \Phi(t)) \cos(2\pi f_0 t) \tag{2.12}$$

To simplify this, recall the identity

$$\cos(\theta_1)\cos(\theta_2) = \frac{1}{2}(\cos(\theta_1 - \theta_2)) + \frac{1}{2}(\cos(\theta_1 + \theta_2)) \tag{2.13}$$

So,

$$s(t) \cdot h(t) = \frac{1}{2}\cos\big(\cancel{2\pi f_0 t} + \Phi(t) - \cancel{2\pi f_0 t}\big) + \frac{1}{2}\cos\big(2\pi f_0 t + \Phi(t) + 2\pi f_0 t\big) \tag{2.14}$$

$$= \frac{1}{2}\cos(\Phi(t)) + \cancelto{0}{\frac{1}{2}\cos(4\pi f_0 t + \Phi(t))} \tag{2.15}$$

The cos() term with $4\pi f_0 t$ is at twice the carrier frequency. The term is canceled when the signal passes through the low-pass filter. This, then leaves

$$I(t) = \frac{1}{2}\cos(\Phi(t)) \tag{2.16}$$

Substituting for $\Phi(t)$, we have

$$I(t) = A_r \,\mathrm{rect}\left(\frac{t - \tau}{T}\right) \cos\big(-2\pi f_0 \tau + \pi K(t - \tau)^2\big) \tag{2.17}$$

Equation (2.17) represents the output of the in-phase channel in our receiver's quadrature demodulation module. Note the $\frac{1}{2}$ is pulled into the constant A_r. This equation is in the form of an LFM chirp at baseband.

A similar set of mathematical steps can be done with the $Q(t)$ channel:

$$Q(t) = A_r \,\text{rect}\left(\frac{t-\tau}{T}\right) \sin\left(2\pi f_0 \tau - \pi K(t-\tau)^2\right) \tag{2.18}$$

These two terms are then sampled with an analog-to-digital converter (ADC) and combined to make a vector of complex numbers. This vector represents the echoes corresponding to one pulse of the radar. Thus, the received signal is

$$s_{qd}[n] = I[n] + jQ[n] \tag{2.19}$$

where n is the index for the digital sampling. Note that because $s_{qd}[n]$ is complex, the Nyquist sampling rate needs to be greater than the bandwidth of the signal, unlike Nyquist sampling for real signals, which requires greater than twice the highest frequency component in the signal.

The exponential form of Equation (2.19) is

$$s_{qd}(t) = A_r \,\text{rect}\left(\frac{t-\tau}{T}\right) e^{j\pi(2f_0\tau - K(t-\tau)^2)} \tag{2.20}$$

For simplicity, these equations will be written as if they are continuous-time functions, despite their digital nature after being sampled by the ADC.

While Equations (2.17) to (2.19) show the demodulated return from one object in the scene, in reality, the received signal will be made up of a collection of scatterers, which provides a superposition of baseband LFM chirps.

To detect each of the baseband, time-delayed LFM chirps, matched filtering will be required. The first term in Equation (2.20) is a constant phase term. The second term, the term we are interested in detecting, is our baseband LFM chirp. The matched filter will need to be of this form.

Recall that a matched filter is a correlation, defined by

$$y(t) = \int_{-\infty}^{\infty} x(u) g^*(u-t)\,du \tag{2.21}$$

where $x()$ is the input signal and $g()$ is the template that is matched. This template is the complex conjugate of the input signal.

In practice, because the vector for a pulse can be large in SAR, matched filtering is nearly always performed in the frequency domain. This is given by

$$Y(f) = X(f)G(f) \tag{2.22}$$

where $X(f)$ is the Fourier transform of $x(t)$ and $G(f)$ is the Fourier transform of $G(f)$.

In the time domain, the matched filter template for our baseband, time-delayed LFM chirp is defined as

$$g(t) = \text{rect}\left(\frac{t}{T}\right) e^{j\pi K t^2} \tag{2.23}$$

The Fourier equivalent of this template is

$$G(f) = \text{rect}\left(\frac{f}{KT}\right) e^{j\pi \frac{f^2}{K}} \tag{2.24}$$

Example 2.1 Consider a radar system with a carrier frequency of $f_0 = 8$ GHz, a chirp rate of $K = 0.6 \times 10^{12}$ Hz/s, and a planned bandwidth of $B = 300$ MHz. From Figure 2.2, it can be shown that the pulse width needed to achieve this bandwidth is given by $B = |K|T$ or $T = B/|K|$. Thus, the pulse width should be $T = 300 \times 10^6$ MHz$/1 \times 10^{12}$ Hz/s $= 50 \times 10^{-6}$ s. Let the round-trip delay time between the sensor and an object be $\tau = 1e^{-6}$ s.

The baseband LFM chirp for the quadrature demodulated received signal takes the form described in Equation (2.20). Figure 2.5 shows the received, demodulated LFM chirp.

The matched filter is then defined as

$$g(t) = w_{\text{mf}}\left(\frac{t}{T}\right) \text{rect}\left(\frac{t}{T}\right) e^{j\pi K t^2} \tag{2.25}$$

where $w_{\text{mf}}()$ is a Kaiser window with $\beta = 2.5$. This windowing function assists in controlling ringing in the output after matched filtering; that ringing is called Gibbs phenomenon.

The time-domain matched filter and its frequency domain response are shown in Figure 2.6. Note that the Fourier transform of an LFM chirp is a rectangle function. The rectangle function is "warped" due to the Kaiser window.

NOTE: It is common practice to take the Discrete Fourier Transform (DFT) using the Fast Fourier Transform (FFT). The FFT is designed so that DC is in the first sample,

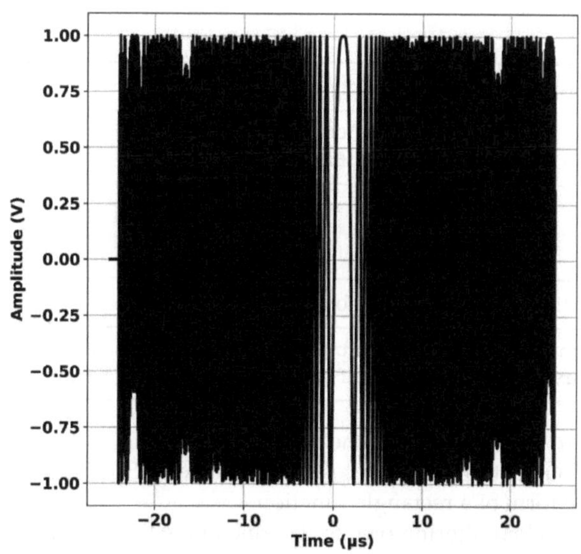

Figure 2.5 The received, quadrature demodulated signal.

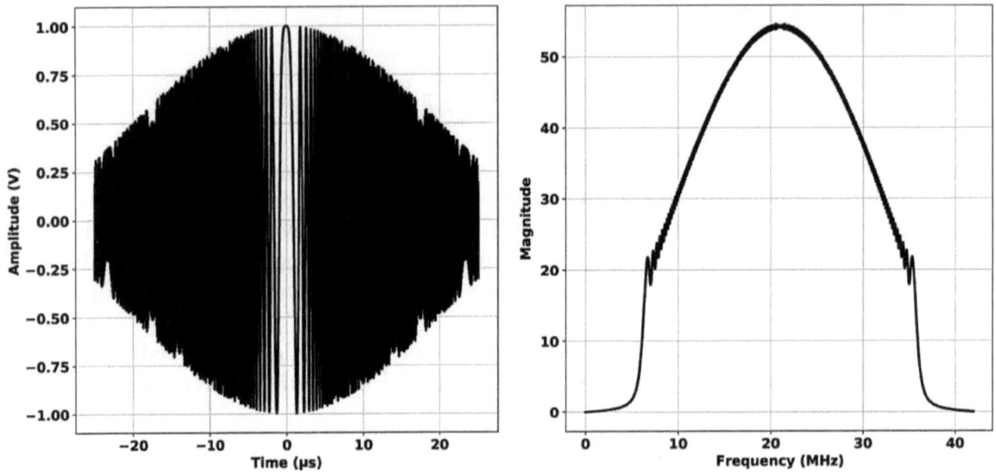

FIGURE 2.6 On the left, the time-domain matched filter, windowed with a Kaiser window with $\beta = 2.5$. On the right, the Fourier transform of the matched filter.

and the left quadrant follows the right quadrant. In MATLAB and Python, there are functions that automatically adjust the signal so that the quadrants are correctly oriented for each FFT input and subsequent output. In both MATLAB and Python, a good practice to reduce errors and keystrokes is to define an anonymous function. Listings 2.1 and 2.2 show the definitions for anonymous functions. The invocation for both the MATLAB and Python functions is `G = fft1d(g);`. The semicolon is typically omitted for Python.

Listing 2.1 MATLAB Anonymous Functions for FFT and IFFT

```
fft1d = @(x) fftshift(fft(fftshift(x)));
ifft1d = @(x) fftshift(ifft(fftshift(x)));
```

Listing 2.2 Python Anonymous Functions for FFT and IFFT

```
fft1d = lambda _x: np.fft.fftshift(np.fft.fft(np.fft.fftshift(_x)))
ifft1d = lambda _x: np.fft.fftshift(np.fft.ifft(np.fft.fftshift(_x)))
```

where "np" is shorthand for the **numpy** package in Python.

The Fourier transform of $s_{qd}(t)$, designated as $S_{qd}(f)$, is shown in Figure 2.7. It is also of the form of a rectangle function.

The matched filtering process is done in the frequency domain

$$S_{\mathrm{mf}}(f) = S_{qd}(f)G(f) \tag{2.26}$$

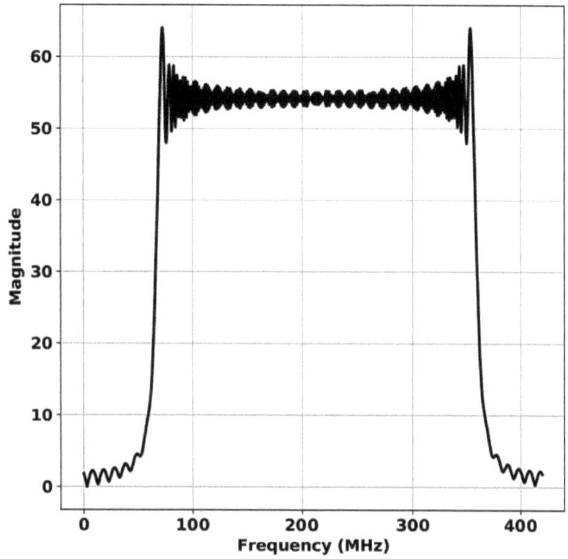

FIGURE 2.7 The Fourier transform of the received, demodulated signal. Note that this is also in the form of a rectangle function.

and an inverse FFT puts the signal back into the time domain

$$s_{\text{mf}}(t) = \mathcal{F}^{-1}(S_{\text{mf}}(f)) \tag{2.27}$$

The resulting matched filter output in the time domain is shown in Figure 2.8.

The resulting output of the matched filtering of an LFM chirp is of the form of a $\text{sinc}^2()$. The main peak is called the *main lobe,* and the peaks to either side of the main lobe are called *side lobes*. The main collection of the main lobe and side lobes is called an *IPR,* which is shorthand for impulse response. Resolution of this IPR is measured at the 3-dB down point, which, when multiplied by the speed of light, gives the resolving distance in the range direction. This is defined as

$$\rho_r = \frac{c}{2B} = \frac{c}{2|K|T} \tag{2.28}$$

Thus, for this example, our resolving distance ρ_r is equal to 5 meters.

2.3.2 Dechirping

Dechirping is similar to quadrature demodulation, but instead of mixing with a sinusoid at the carrier frequency, a time-delayed copy of the chirp is used. When this is combined with a low-pass filter to eliminate the frequencies above the carrier frequency, what is left is

$$s_{dc}(t) = A_r \,\text{rect}\left(\frac{t-(\tau-\tau_c)}{T}\right) e^{j\left[2\pi K(\tau-\tau_c)t + 2\pi f_0(\tau-\tau_c) - \pi K(\tau^2-\tau_c^2)\right]} \tag{2.29}$$

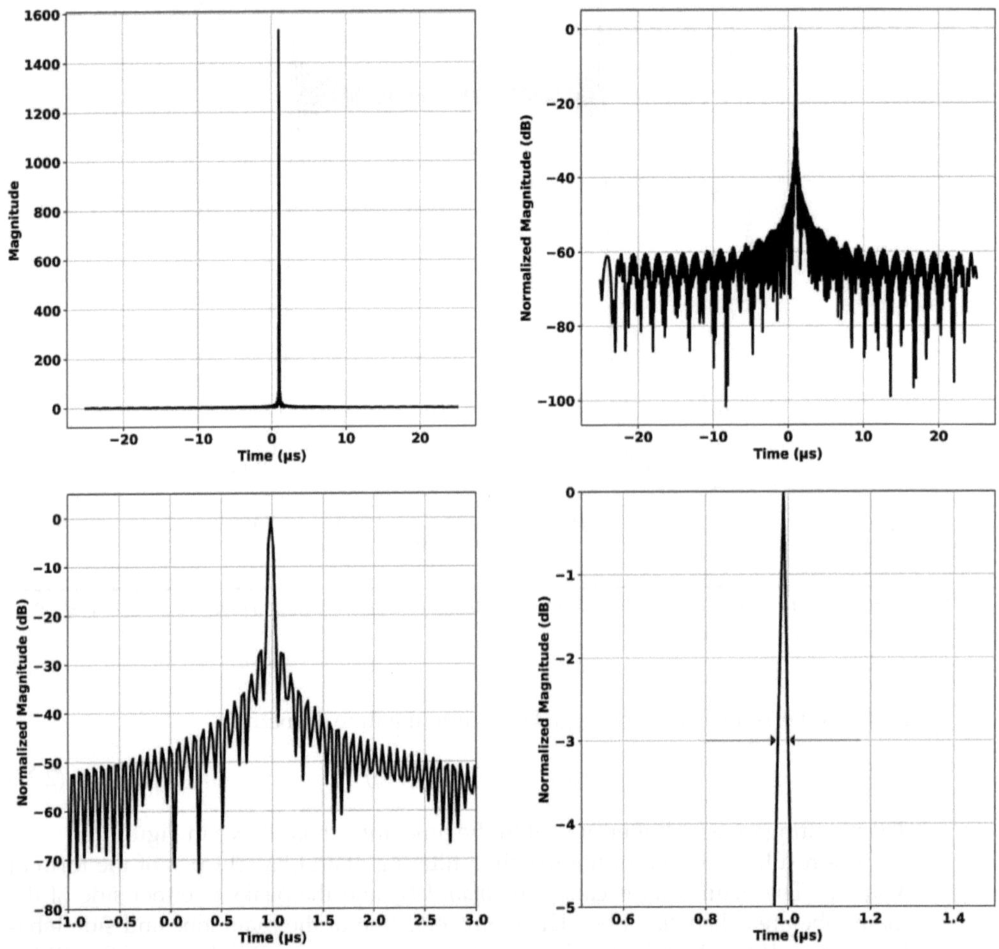

FIGURE 2.8 The matched filter output in the time domain with magnitude (top left), magnitude in decibles (top right), magnitude in decibles zoomed into main lob (bottom left), and magnitude in decibles zoomed into the 3-dB down area (bottom right).

where τ_c is the round-trip time delay to the center of the scene, τ is the round-trip time delay to some arbitrary object in the scene, T is the width of the pulse, and f_0 is the carrier frequency. This result is in the form of a single sinusoid with a frequency of $K(\tau - \tau_c)$ Hz. Digital matched filtering after the ADCs is not necessary because the dechirp inherently mixes out the baseband LFM chirp, leaving only a single sinusoid. That sinusoid correlates directly with an object at some distance $R = \frac{c\tau}{2}$.

Figure 2.9 shows the processing flow for dechirp. Note the similarities to quadrature demodulation as shown in Figure 2.4. The cos and sin are replaced with replicas of the transmitted chirps, delayed by τ_c, which corresponds to the round-trip time delay to the GRP. As has been mentioned, it's common to reference to a specific spot (in the case of

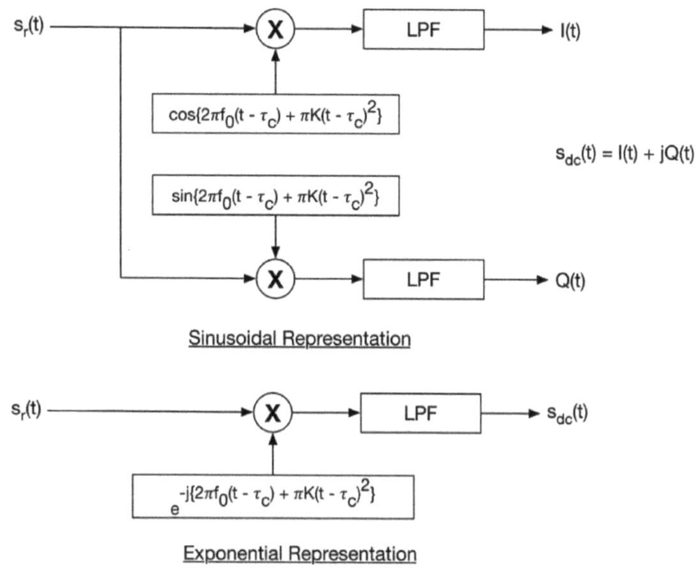

FIGURE 2.9 The flowchart for dechirping. Note the similiarities with quadrature demodulation, as shown in Figure 2.4.

spotlight imaging) or line (in the case of stripmap imaging). For the quadrature channel (Q channel), the chirp is phase shifted by 90° to provide coherency.

To further illustrate how dechirp works, consider Figure 2.10. Note that this illustration has an exaggerated scale to allow for effective visualization.

To start the process, a chirp of width $\tau_p = T$ is transmitted. The round-trip delay time from the nearest edge to the furthest edge determines when to start listening and when to stop listening before the next pulse is transmitted. The time delay for the wave to travel across the scene is τ_D. The width of the pulse τ_p must be larger than the time for the signal to transverse across the scene τ_D. The chirp rate K determines the slope of the LFM chirp and ultimately provides for the transmitted bandwidth B

$$B = |K|T = |K|\tau_p \tag{2.30}$$

If there are two objects in the scene, one between the near edge and the GRP and the other between the GRP and the far edge, the round-trip time delay for those objects is τ_1 and τ_2, respectively. Just as before, the first return seen by the radar will come from Target 1 and will be an attenuated, time-delayed copy of the transmitted LFM chirp. The second return will then come from the second target and will also be attenuated (but likely not attenuated exactly as Target 1) and time-delayed.

Mixing and low-pass filtering are similar to what's described in Section 2.3.1, the details of which are left to the reader. What will be left after this processing is a superposition of two sinusoids, one corresponding to Target 1 and the other corresponding to Target 2. A Fourier transform of $s_{dc}(t)$ will compress the signal, providing an IPR for each target. The bandwidth needed to represent the dechirped targets, called the *video bandwidth* B_v, is much smaller than the transmitted bandwidth B.

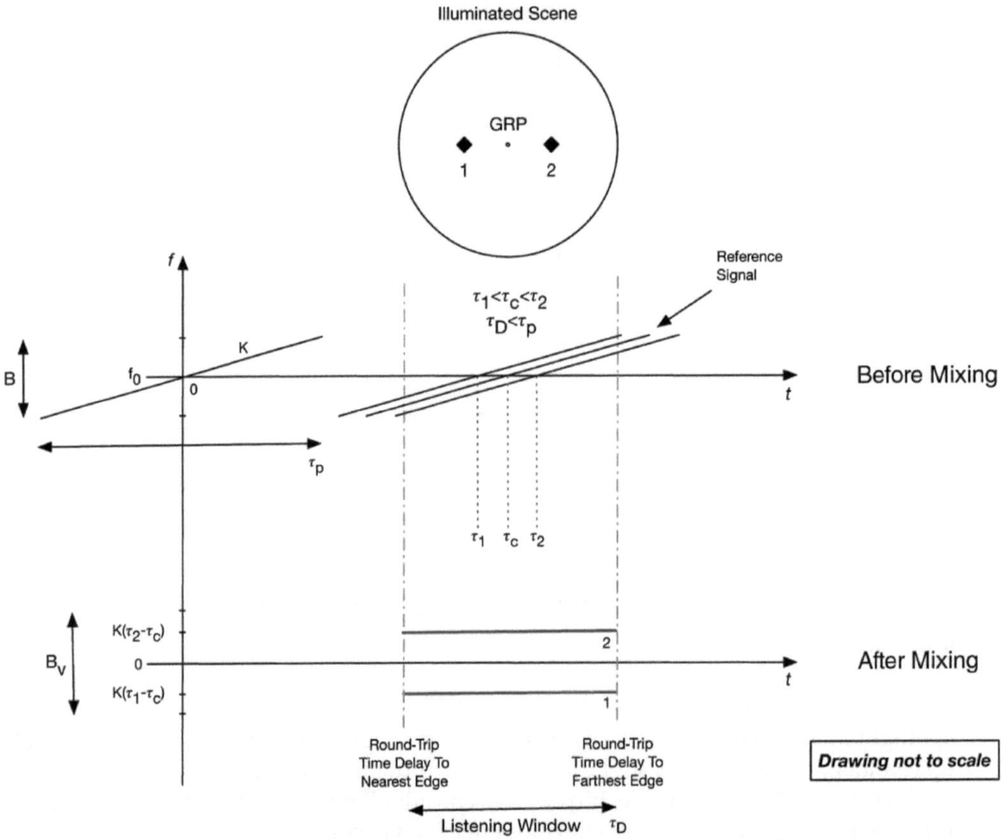

Figure 2.10 Graphical illustration of dechirping, considering what happens to the signals in the frequency domain.

Example 2.2 Consider the radar system described in Example 2.1. Let the round-trip time delay to the GRP be 2 μs. After dechirp, the sinusoid that corresponds to the object at $\tau = 1$ μs is

$$\begin{aligned} f &= K(\tau - \tau_c) \\ &= 0.6 \times 10^{12}\ \text{Hz/s} \cdot (1 \times 10^{-6} - 2 \times 10^{-6})\ \text{s} \\ &= -600{,}000\ \text{Hz} \\ &= -600\ \text{MHz} \end{aligned} \tag{2.31}$$

An object at the scene center would be at DC, while an object at 3 μs would be at +600 MHz.

There are advantages and disadvantages in using dechirp instead of quadrature demodulation. One advantage is the elimination of the matched filtering step after demodulation. Another is the reduction of bandwidth after demodulation, representing a reduced ADC sampling rate and reduced data storage. The biggest disadvantage is the relative complexity of the dechirp processing chain.

Now that ranging has been described, the next chapter will examine how an ensemble of these pulses can be employed to separate targets in the along-track direction. After the raw data has been collected, typically called *video phase history* (VPH) data for historical reasons, more processing is needed to form a quality image that can be viewed or exploited automatically.

CHAPTER 3
Imaging

In this chapter, we'll explore the mathematical models that describe how synthetic aperture radar (SAR) imaging works. The models will consider the similarities and differences in the quadrature demodulated and dechirped data. The chapter concludes by stepping through a full simulation for a stripmap collection mode using quadrature demodulation, as well as a spotlight collection mode using dechirp.

3.1 Mathematical Model

Now that we've constructed the model for a single ranging pulse, let's look at pulses over time. Recall from Chapter 1 that there is a pulse train transmitted from the radar.[1] In between each transmitted pulse is a listening window, a period of time where the radar listens for reflected returns. Reflected returns are demodulated and sampled; those vectors of sampled data are then stored in a matrix, indexed by their "slow time" index. Figure 3.1 illustrates the simple relationship between fast and slow times. Recall that fast time t is the dimension and scale for each individual listening window. *Slow time* η is the dimension and scale for the ensemble of sampled pulses.

The position of the radar is changing over the course of slow time, by $y_a(\eta) = v_a\eta + y_0$, where y_a is the current position of the aircraft or satellite, v_a is its actual velocity, and y_0 is its initial position at the start of the dwell. This assumes the path of the aircraft or satellite is linear with constant velocity; this is rarely the case for an aircraft and never the case for a satellite. The actual position, along with any kinematic information, is recorded by a Global Position Satellite (GPS) system and Inertial Navigation System (INS). This information is then used during image formation processing to correct the phase error associated with a nonlinear, nonconstant velocity flight track.

Refer back to Figures 1.9 and 1.10. Let us assume that the objects in the scene are stationary. With that in mind, the range over slow time changes for each object in the scene uniquely. The mathematical proof for this is simple and is left to the reader. That change in range always takes on the form of a paraboloid. Consider the Euclidean distance between the sensor and an arbitrary object in the scene

$$R(\eta) = \sqrt{(x_{\text{obj}} - x_a)^2 + (y_{\text{obj}} - y_a(\eta))^2 + (z_{\text{obj}} - z_a)^2} \tag{3.1}$$

[1] Note that this is true for a pulsed radar system; while this is still conceptually true for a linear frequency-modulated continuous wave (LFMCW) system, the implementation of such a system is different than a pulsed SAR system. This will be covered in detail in a later chapter.

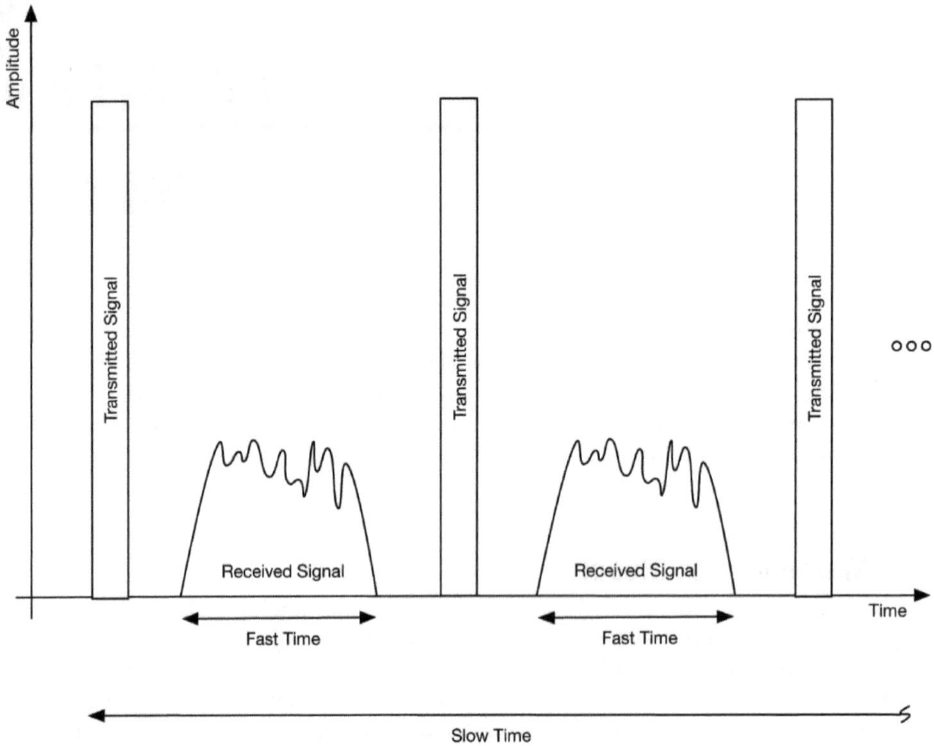

FIGURE 3.1 An illustration of fast and slow times during a SAR collect.

where $(x_{\text{obj}}, y_{\text{obj}}, z_{\text{obj}})$ is the stationary position of the object and $(x_a, y_a(\eta), z_a)$ is the position of the aircraft or satellite. Note that if we are assuming a straight and level flight, only y_a is changing over η. Since $y_a(\eta)$ is the only variable changing in Equation (3.1), this creates a unique parabola for each object in the scene, as shown in Figure 1.15. The minimum of this parabola is the closest point of approach (CPA).

Leveraging Equations (2.20) and (2.29), the raw radar data over the course of the dwell is defined as

$$s_{\text{qd}}(\eta, t) = A_r w_a(\eta) \,\text{rect}\left(\frac{t - \tau(\eta)}{T}\right) e^{j\pi(2f_0\tau(\eta) - K(t-\tau(\eta))^2)} \tag{3.2}$$

for quadrature demodulated data and

$$s_{\text{dc}}(\eta, t) = A_r w_a(\eta) \,\text{rect}\left(\frac{t - (\tau(\eta) - \tau_c)}{T}\right) e^{j\left[2\pi K(\tau(\eta)-\tau_c)t + 2\pi f_0(\tau(\eta)-\tau_c) - \pi K(\tau(\eta)^2 - \tau_c^2)\right]} \tag{3.3}$$

for dechirped data. In both equations, $w_a(\eta)$ describes the antenna beam pattern in azimuth. This factor is dependent on the type of antenna or antennas (some designs may use separate transmit and receive antennas, while other systems may use only one and toggle between transmit and receive).

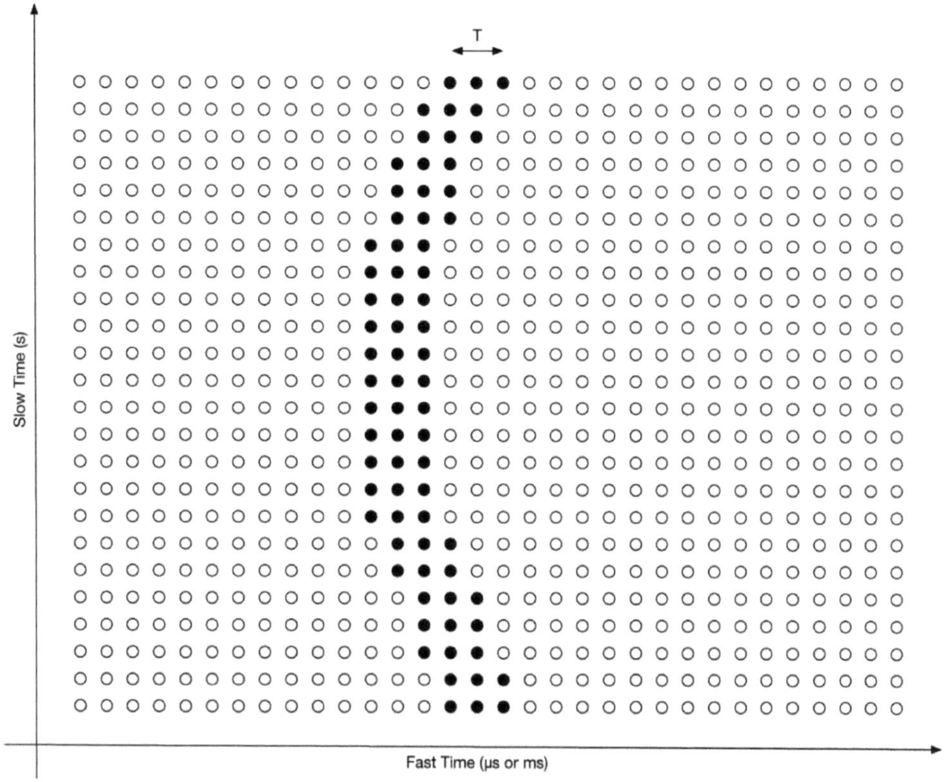

FIGURE 3.2 The slow and fast time grid with a single arbitrary target. The width of the signal T corresponds to the transmitted pulse width.

In reality, the data has already been sampled by the analog-to-digital converter, and each η is actually an index to a vector. The ensemble of vectors is stored in a matrix and represents the unfocused SAR image. *Image formation* is a set of algorithms that can be used to focus the raw radar data, also known as video phase history (VPH) data. Figure 3.2 shows a simple sample grid with a single return. In fast time, the signal is T seconds in width, corresponding to the width of the transmitted pulse. This corresponds to the rect function in Equations (3.2) and (3.3). The aforementioned parabola is visible over slow time, due to the change in the time delay between the object and the sensor.

Noting that $\tau = \frac{2R}{c}$, Equations (3.2) and (3.3) can be referenced from their ranges by

$$s_{\text{qd}}(\eta, t) = A_r w_a(\eta)\,\text{rect}\left(\frac{t - \left(\frac{2R(\eta)}{c}\right)}{T}\right) e^{j\pi\left[2f_0\left(\frac{2R(\eta)}{c}\right) - K\left(t - \left(\frac{2R(\eta)}{c}\right)\right)^2\right]} \tag{3.4}$$

and,

$$s_{\mathrm{dc}}(\eta, t) = A_r w_a(\eta)\, \mathrm{rect}\left(\frac{t - \left(\frac{2R(\eta)}{c} - \tau_c\right)}{T}\right) e^{j\left[2\pi K\left(\frac{2R(\eta)}{c} - \tau_c\right)t + 2\pi f_0\left(\frac{2R(\eta)}{c} - \tau_c\right) - \pi K\left(\left(\frac{2R(\eta)}{c}\right)^2 - \tau_c^2\right)\right]} \tag{3.5}$$

3.1.1 The Doppler Effect

Like most other texts on radar, we have been treating the system as a "stop-and-hop" system, where the position of the radar is used to understand the range between the sensor and an object, but the actual motion of the system is not taken into consideration. Because the aircraft or spacecraft is in motion, there is a Doppler effect induced on each pulse. That effect changes the received frequency either higher or lower, depending on the relative change in range to the object. This is directly analogous to the change in pitch of a train whistle. If the train were stationary, the whistle would emit a pitch of a single frequency. As the train moves toward you, the whistle's pitch sounds higher, while as it moves away from you, the whistle's pitch sounds lower. This is also demonstrated in light with "blue-shifted" and "red-shifted" light, which corresponds to objects moving toward us and objects moving away from us, respectively.

The Doppler effect within a received pulse is negligible, even for fast moving spacecraft, because of the relatively short period of time that it takes to send out a pulse and listen for its return. This period of time, called the *pulse repetition interval* (PRI), is so short that it gives the appearance that the range between the sensor and the object is static within that interval. Because of this, the Doppler effect in fast time is safely ignored and the "stop-and-hop" model can be used. This is not the case for some linear frequency-modulated continuous wave (LFMCW) systems, as will be discussed in Chapter 9.

In slow time, there is a change in frequency due to the change in range $R(\eta)$. This change in range induces a change in frequency due to our radar processing. Depending on whether an up chirp or a down chirp was transmitted, the received frequency will either increase or decrease depending whether the sensor is moving toward an object or away from an object. This is why the slow-time dimension is often referred to as Doppler or Doppler shift. The average Doppler shift is called the *Doppler centroid*.

Video phase history (VPH) data lives in the (η, t) slow-time, fast-time space. When each pulse is matched filtered and a Fast Fourier Transform (FFT) is applied, this puts the data in that pulse in range space. If done for all the collected pulses, this is called the *range-Doppler* space. If further processing is performed and a matched filter and FFT are applied to the Doppler vectors, then the data is transformed into *range-azimuth* or *image* space. In this space, if the processing is done properly, the focused SAR image is "detected" by taking the magnitude of the complex samples.

3.1.2 Pulse Repetition Frequency and Doppler Chirp Rate

For both quadrature demodulated and dechirped data, we have a slowly varying linear frequency-modulated (LFM) chirp across slow time. Like any other signal, we must meet the Nyquist requirement to avoid aliasing. The rate at which we sample is dictated by the *pulse repetition frequency* (PRF), which is the frequency of transmitted pulses and their respective listening windows. The *PRI*, which is the reciprocal of the PRF, is the time period between the start of the pulse and the start of the next pulse, which includes the time for the listening window.

The selection of the PRF is more an art form than an exact science. The minimum frequency is bounded by the Doppler bandwidth,

$$\Delta f_d = \frac{2v_a \cos\theta}{\lambda}\beta_a \tag{3.6}$$

where θ is the squint angle and β_a is the 3-dB width of the radar beam in azimuth. That beamwidth is defined as

$$\beta_a = \frac{0.886\lambda}{L_a} \tag{3.7}$$

where L_a is the physical width of the antenna in azimuth. Thus, the Doppler bandwidth can be written as

$$\Delta f_d = \frac{2v_a \cos\theta}{\lambda}\frac{0.886\lambda}{L_a} = \frac{1.77v_a \cos\theta}{L_a} \tag{3.8}$$

The Doppler bandwidth sets the lower bound for the PRF; that is, the PRF may not be lower than Δf_d, else aliasing may occur. Typically, the PRF is set to the Doppler bandwidth times some oversampling frequency, typically in the 1.1 to 1.4 range.[2]

Example 3.1 Consider an X-band airborne SAR system, traveling at a constant velocity of 125 m/s, with a physical antenna width of 0.1 m, and a center wavelength of 3 cm. The antenna is pointing broadside $\theta = 0$. The 3-dB azimuth beamwidth for this system is

$$\beta_a = \frac{0.886(0.03\text{ m})}{0.1\text{ m}} = 0.266\text{ rad} = 15.23^\circ \tag{3.9}$$

The Doppler bandwidth for this system and collection geometry is

$$\Delta f_d = \frac{2(125\text{ m/s})\cos(0)}{0.03\text{ m}}\left(0.266\text{ rad}\right) = 2217\text{ Hz} = 2.2\text{ kHz} \tag{3.10}$$

If the oversampling factor O is 1.2, then the PRF is defined as

$$\text{PRF} = O \cdot \Delta f_d = (1.2)(2217\text{ Hz}) = 2660\text{ Hz} = 2.66\text{ kHz} \tag{3.11}$$

The *Doppler chirp rate* (also called the azimuth FM rate) is the rate of change of the Doppler frequency over slow time. It is defined as

$$K_a = \frac{2}{\lambda}\frac{d^2R(\eta)}{d\eta^2}\bigg|_{\eta=\eta_c} \tag{3.12}$$

$$= \frac{2v_a^2 \cos^2\theta}{\lambda R(\eta_c)} \tag{3.13}$$

---

[2] I. G. Cumming and F. H. Wong, *Digital Processing of Synthetic Aperture Radar Data: Algorithms and Implementation*, Artech House, Boston, MA, pp. 140–141, 2005.

where η_c is the scene center time and $R(\eta_c)$ is the slant range to the scene center. The scene center time is defined as the time for the beam to reach the scene center for stripmap mode and halfway through the dwell for spotlight mode.

3.1.3 Synthetic Aperture and Exposure Time

A real aperture radar system's azimuth resolution is determined by the azimuth beamwidth. The beamwidth is influenced by the antenna length L_a and the radar wavelength λ. Because both antenna length and wavelength are fixed for a given radar system, the only way to create a higher resolution is to effectively reduce the azimuth beamwidth. This is done by using signal processing to create a narrower beam in azimuth. A similar technique was used to "compress" the range pulse. The antenna aperture is inversely proportional to the beamwidth. Because of this, synthesizing a smaller beamwidth means creating a larger aperture. Figure 3.3 illustrates the antenna azimuth beamwidth and the synthetic aperture/angle. This illustration does not consider any significant squint, but squint plays a factor in the length of the synthetic aperture. The diagram shows the azimuth beamwidth θ_s and the antenna beamwidth β_a. These angles are related to the velocity of the aircraft,

$$\theta_s = \frac{v_a}{v_g}\beta_a \tag{3.14}$$

where v_a is the velocity of the aircraft along the flight path and v_g is the velocity of the aircraft projected into the ground plane. The synthetic aperture is described in Figure 3.3

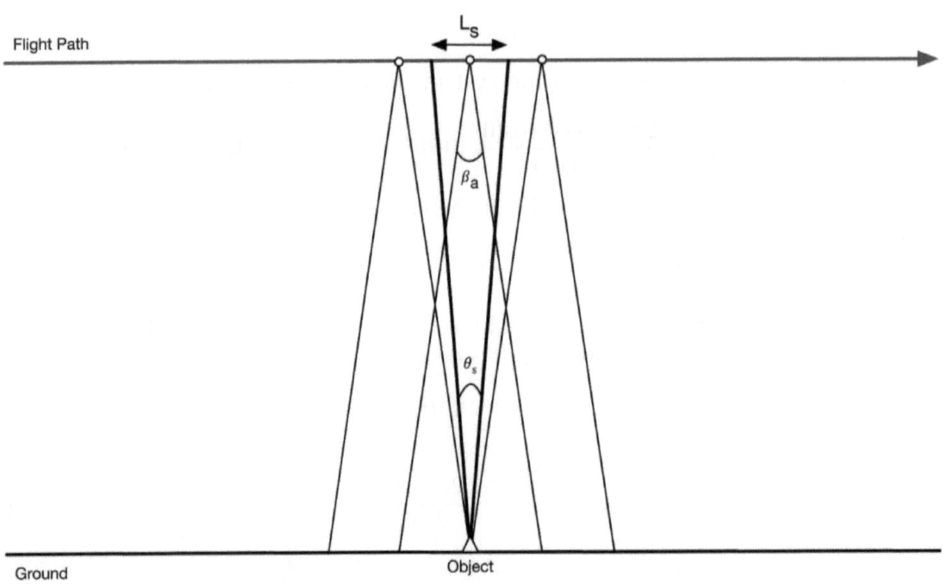

FIGURE 3.3 A diagram illustrating synthetic aperture for a stripmap mode collection.

by L_s. It is given by

$$L_s = \frac{0.886 R_0 \lambda}{L_a \cos(\theta)} \frac{v_a}{v_g} = \frac{0.886 R(\eta_c)\lambda}{L_a} \tag{3.15}$$

where the constant 0.886 is a 3-dB beamwidth broadening factor attributed to the antenna beam pattern, R_0 is the slant range of closest approach, $R(\eta_c)$ is the range at zero Doppler, and θ is the squint angle. The synthetic aperture for aircraft systems can be tens to hundreds of meters long, depending on the aircraft's physical antenna size and the velocity. Satellite SAR systems can require synthetic apertures on the order of kilometers.

The time it takes the azimuth 3-dB beamwidth beam pattern to transverse over an object in the scene is called the *exposure time*. For a spotlight system, because the target is illuminated over the course of the entire dwell, the exposure time is equivalent to the total slow time, or $T_a = \eta_{tot}$, where T_a is the object exposure time.

For a stripmap system, this is the time it takes for the azimuth antenna beam pattern to transverse across the object (from start to finish). Mathematically, this is defined as

$$T_a = \frac{0.886 V \cos\theta}{\Delta f_d} \tag{3.16}$$

In a spotlight system, the exposure time is equivalent to the total dwell time, as the target is always in the beam during imaging.

3.1.4 Azimuth Resolution

As was discussed in the previous section, the azimuth resolution is tied directly to the azimuth beamwidth. In stripmap imaging, it's dependent on the azimuth beamwidth, while for spotlight, it's dependent on the total squint angle during the dwell. It was also demonstrated that the return signal across pulses is frequency modulated. This is due to the motion of the platform. The azimuth frequency modulation is at much lower intermediate frequencies than the range frequency modulation. Thus, matched filtering in both the range and azimuth directions will compress the signal and provide the high resolution IPRs needed.

Just as range resolution is dependent on the radar bandwidth, azimuth resolution is dependent on the Doppler bandwidth. This is defined as

$$\rho_a = \frac{0.886 v_g \cos\theta}{\Delta f_d} = \frac{L_a}{2} \frac{v_g}{v_a} = \frac{0.886\lambda}{2\theta_s} \tag{3.17}$$

Example 3.2 Consider an X-band synthetic aperture radar system with a wavelength of $\lambda = 0.03$ meters, a ground velocity of $v_g = 200$ meters per second, a squint angle of $\theta = 60$ degrees, and a Doppler bandwidth of $\Delta f_d = 450$ Hz. The expected azimuth resolution of such a system would be

$$\rho_a = \frac{0.886(200 \cdot \cos(60^\circ))}{450} = 0.2 \text{ m} \tag{3.18}$$

Example 3.3 Consider an X-band synthetic aperture radar system with a wavelength of $\lambda = 0.03$ meters, with a physical antenna length of $L_a = 2$ meters. Assuming the ground and flight velocities are equal, the expected azimuth resolution of such a system is

$$\rho_a = \frac{L_a}{2} = \frac{2}{2} = 1\ \text{m} \tag{3.19}$$

If it is desired that range resolution matches azimuth resolution, then the radar bandwidth must be

$$B = \frac{c}{2\rho_r} = \frac{299792458\ \text{m/s}}{2 \cdot 1\ \text{m}} \approx 150\ \text{MHz} \tag{3.20}$$

3.2 Simulated Point Scatterers Using Quadrature Demodulation

To illustrate the concepts in this and the previous chapter, let us consider a hypothetical radar system, whose parameters are shown in Table 3.1. To simplify, we will consider the SAR in two dimensions (eliminating z as a variable). Range is then calculated with x and y, using Euclidean distance. In practice, this is not practical, but makes simulating the system a little easier.

Consider three ideal point scatterers, as shown in Figure 3.4; the graph shows the targets with respect to the scene center (also known as the ground reference point or GRP, displayed as a plus or cross symbol at the center of the graph). The sensor is on the left-hand side of the plot, moving from bottom to top. Targets B and C are in line with one another in the azimuth direction, but separated in range. Targets A and C are in line with one another in the range direction, but separated in azimuth. The iso-range lines for the radar are idealized as vertical lines at each range cell. Thus, in this example, a

Parameter	Variable	Value
Slant range from aircraft to scene center at zero doppler	$R(\eta_c)$	15 km
Effective radar velocity	V_a	150 m/s
Transmitted pulse duration	T_r	2.5 μs
Range FM rate	K_r	20 MHz/μs
Radar center frequency	f_0	9.8 GHz
Doppler bandwidth	Δf_d	78 Hz
Range A/D sampling rate	F_r	60 MHz
Pulse repetition frequency	F_a	100 Hz
Number of azimuth samples	N_a	256
Number of range samples	N_r	320
Squint angle	θ	0° (broadside)
Beam center cross time	η_c	0 s
Doppler centroid frequency	f_{η_c}	320 Hz

TABLE 3.1 System Parameters for Hypothetical X-Band Synthetic Aperture Radar

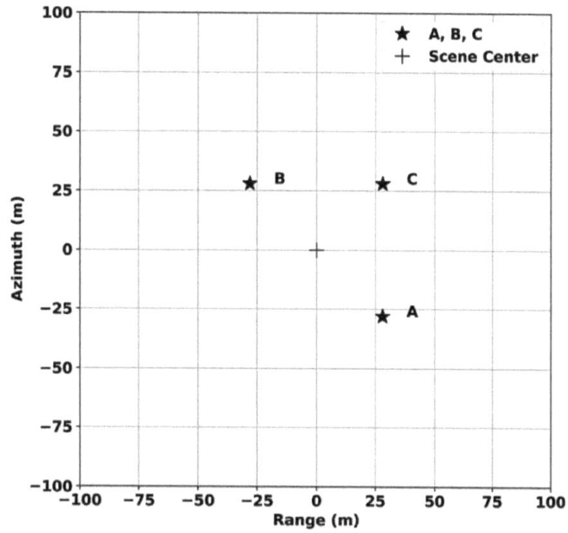

FIGURE 3.4 Three ideal point scatterers in an idealized 200 × 200 meter scene.

real radar would be able to distinguish Target B from A and C, but would not be able to distinguish A and C from one another.

Knowing that the system will take 320 pulses, the total dwell time η_{tot} can be calculated as

$$\eta_{\text{tot}} = N_a \frac{1}{F_r} = 256 \frac{1}{100 \text{ Hz}} = 2.56 \text{ s} \tag{3.21}$$

Thus, the total distance traveled by the aircraft during the dwell is

$$L_s = \eta_{\text{tot}} v_a = (2.56 \text{ s})(150 \text{ m/s}) = 384 \text{ m} \tag{3.22}$$

The range to the targets can be shown in Figure 3.5. Note that scatterers A and C appear to have nearly the same ranges. The zoom in plot (right side of figure) shows that they do, in fact, have unique ranges over slow time. As we will see, it's this uniqueness in range and our ability to match filter for it, that will separate the targets and allow us to create the reflectivity map of the scene being imaged.

The next aspect to consider is the antenna beam pattern. Each antenna will have a beam pattern that is commensurate with its type. We will consider an antenna model that uses the following model for the illumination beam pattern:

$$w_a(\eta) = \text{sinc}\left(\left(\frac{L_a}{\lambda}\right) \tan^{-1}\left[v_a\left(\eta - \left(\eta_c + y_a(\eta)/v_a\right)\right) / \left(R_{\eta_c} + x_a(\eta)\right)\right]\right)^2 \tag{3.23}$$

where η_c is the time when the sensor arrives at zero Doppler, x_a and y_a are the position of the aircraft, v_a is the velocity of the aircraft, R_{η_c} is the range between the sensor and the scene center at η_c, L_a is the physical length of the antenna (on the order of 1 to 5 meters), and λ is the center wavelength of the radar. Figure 3.6 shows the antenna illumination pattern, as defined in Equation (3.23), for all three scatterers. The illumination pattern

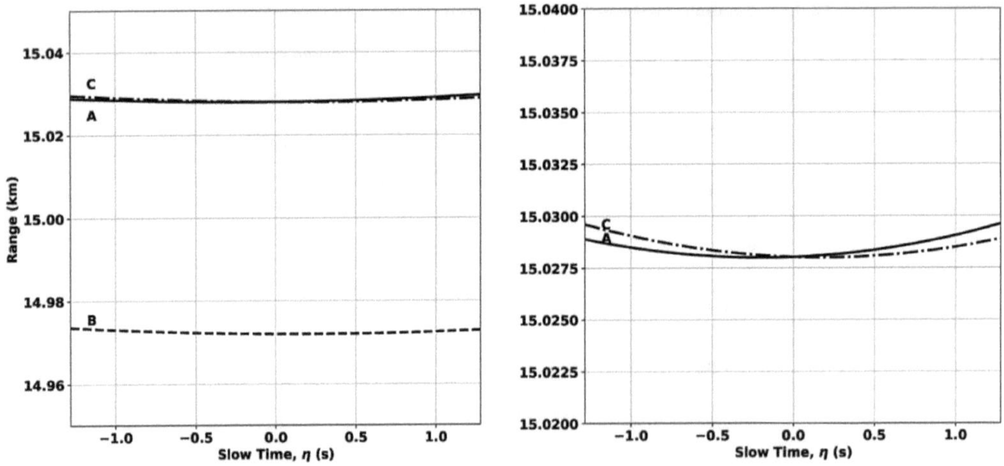

Figure 3.5 The range from the sensor to each individual scatterer over slow time. Plot on the left is all three scatterers. Plot on the right is a zoom-in of scatterers A and C.

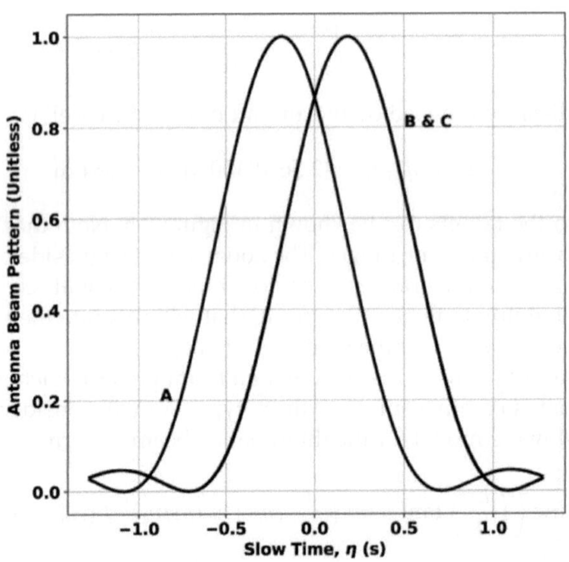

Figure 3.6 The antenna illumination pattern over slow time as it hits each of the targets. Because B and C are in the same azimuth bin, their illumination patterns are equal.

over slow time is a sinc-squared function. Because scatterers B and C are in the same azimuth bin, their illumination patterns are equal.

The range and azimuth resolutions for this system and mode of collection are

$$\rho_r = \frac{c}{2}\frac{1}{KT} = 3\text{ m} \tag{3.24}$$

where c is the speed of light, K is the chirp rate, and T is the pulse width, and,

$$\rho_a = \frac{L_a}{2} = 1.7\text{ m} \tag{3.25}$$

where L_a is the physical width of the antenna. The azimuth resolution is approximately 2 times less than the range resolution. Once the data is collected (or, in this case, simulated), the data can be down sampled to match the resolutions.

Using Equation (3.2), each pulse can be mathematically simulated. Figure 3.7 shows the results. From top left to bottom right, the plots represent the real part of s_{qd}, the imaginary part of s_{qd}, the magnitude of s_{qd}, and, finally, the phase of s_{qd}. The "X" shape is the response from one of the scatterers. The lone X shape is from scatterer A, while scatterers B and C have their phases intertwined as they are in the same azimuth bin. A useful way to mentally visualize these plots is to recall that the sensor is moving from bottom to top. Pulses run from left to right and each pulse is stacked one on top of the other. That stacking makes the pattern in each of the plots in Figure 3.7. As the sensor moves upward, the antenna beam, which is pointed to the right at broadside (i.e., with no squint angle), encounters scatterer A first (refer back to Figure 3.4). As the beam transverses, it lights up A and then dims on A. As it passes the point of closest approach with A, the antenna

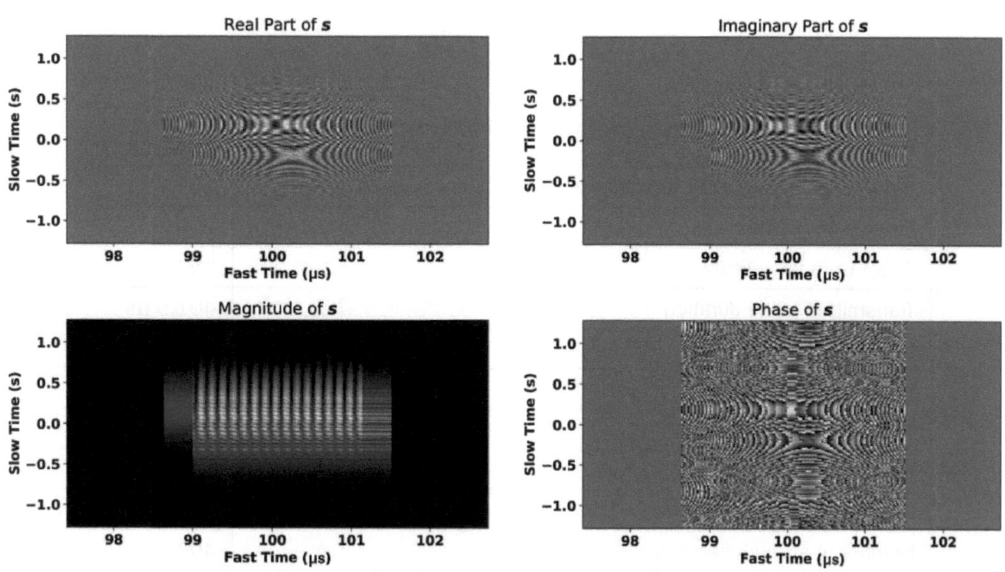

Figure 3.7 The results from simulating three scatterers in a stripmap configuration using quadrature demodulation. From top left to bottom right, the plots show the real, imaginary, magnitude, and phase of s_{qd}.

beam illuminates B and C. It continues to dim on A, as illustrated in Figure 3.6. It then reaches closest approach with B and C, then dims until the end of the collection.

Now that we have collected data on our scene, a natural question is "how will an image be formed from this?" This question will be explored in the next couple of chapters. This data, that is, the data described by s_{qd}, goes by many names in other sources: VPH, phase history, raw data, etc. In this book, we will refer to *raw data* is data before quadrature demoduation or dechirp, and *unfocused phase data* as the data described by s_{qd} or s_{dc} [Equation (3.4) or (3.5)].

3.3 Simulated Point Scatterers Using Dechirp

The simulation of point scatterers using dechirp is somewhat different from the previous simulation. They are shown in Table 3.2. Here we will assume a spotlight mode geometry, where the antenna beam will be pointed to the same spot on the ground. The slewing of the antenna to maintain the same spot on the ground is either done electronically or mechanically. Because of this, the antenna beam pattern isn't as important and will be ignored in this example for the sake of simplicity.

Many of the same parameters from the previous example will be used for this example. However, for dechirp to be effective, the pulse duration must be longer than the time it takes for the LFM chirp to travel across the patch. That is, the pulse width may not be shorter than the total round-trip time across the patch or else the demodulation that occurs using the reference pulse will not match up temporally with the return signals from each scatterer in the scene. This can be seen in Figure 2.10 and is captured in the equation

$$T_{\text{eff}} = T - (2D/c) \tag{3.26}$$

Parameter	Variable	Value
Slant range from aircraft to scene center at center of dwell	$R(\eta_c)$	15 km
Patch/swath width	D	200 m
Effective radar velocity	V_a	150 m/s
Transmitted pulse duration	T_r	401.5 μs
Effective pulse duration	T_{eff}	400 μs
Range FM rate	K_r	1 MHz/μs
Radar center frequency	f_0	9.8 GHz
Range A/D sampling rate	F_r	1 MHz
Pulse repetition frequency	F_a	163.5 Hz
Number of azimuth samples	N_a	400
Number of range samples	N_r	400
Squint angle (through whole dwell)	θ_s	−1.34° to 1.34°

Table 3.2 System Parameters for Hypothetical X-Band Synthetic Aperture Radar

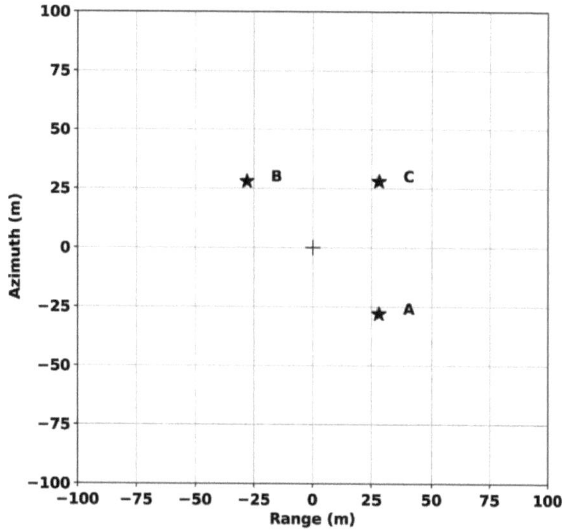

FIGURE 3.8 The placement of three ideal point scatterers in the scene for this example.

where T_{eff} is called the *effective pulse width time*, T is the actual pulse width, and D is the patch/swath width. In this example, the pulse width is 401.5 μs, so the effective pulse width is

$$T_{\text{eff}} = 401.5 \times 10^{-6}\,\text{s} - (2 \cdot 200\,\text{m}/3 \times 10^{8}\,\text{m/s}) = 400\,\mu\text{s} \tag{3.27}$$

That is, it takes the linear FM chirp 1.5 μs to travel across the swath (in the range direction) and back.

Consider three targets, shown in Figure 3.8, placed in the same configuration as the previous example, with the aircraft traveling from the bottom of the graph to the top along the y-axis. Since the desired squint angle through the dwell ranges from $-1.34°$ to $1.34°$, this gives a total of $2.68°$ squint. This can be seen in Figure 3.9. Using trigonometry, we can calculate the length of the synthetic aperture to be 701.7 meters. From here, we can calculate the total imaging or dwell time to be 4.7 seconds. The range resolution is calculated using the same relationship as Equation (3.24), which comes to be 0.37 meters. However, azimuth resolution is calculated using the total squint:

$$\rho_a = \lambda/2\theta_s = 0.03\,\text{m}/2(0.05\,\text{rad}) = 0.32\,\text{m} \tag{3.28}$$

This implies that our impulse response (IPR) should be approximately symmetric in both azimuth and range.

The range from the sensor to the ground reference point (GRP) is the same as in the previous example. Given this, Equation (3.5) is used to simulate the return signal from the three scatterers over the dwell. Again note that the antenna beam pattern term w_a is ignored for this example, as the beam is staring at the same spot for all time. There will be signal roll-off at the start and end of each returned pulse, but for the sake of simplicity, we'll ignore that in this example.

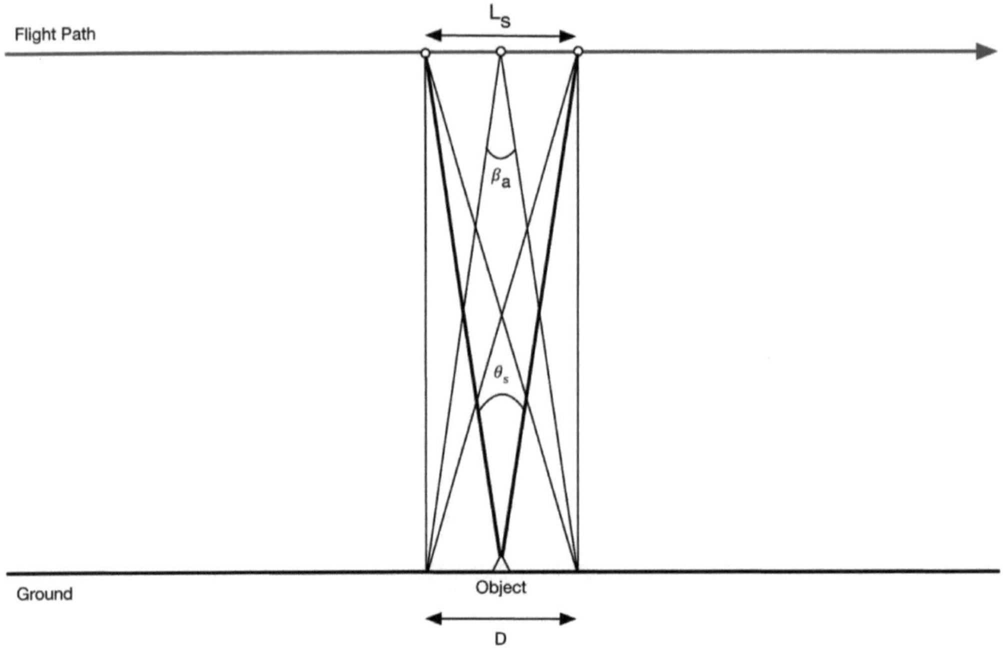

FIGURE 3.9 A diagram illustrating the synthetic aperture for a spotlight mode collection.

Recall that dechirp mixes the return signal with a reference chirp and low-pass filters it, creating a sinusoid in fast time for each scatterer present in the scene. All the pulses are collected by calculating the position of the aircraft with respect to each of the scatterers and creating the superposition of sinusoids. For the three targets in our example, the magnitude of the resulting dechirped signal is shown in Figure 3.10.

Note that because we have restricted the range sampling to the width of the swath and not the pulse width (as we did in the previous example), there is no zero-padded spaces before and after each pulse as shown in Figure 3.7. Given the form of Equation (3.5), the results represent unique sinusoids for each target in the form of the term $2\pi K\left(\frac{2R(\eta)}{c} - \tau_c\right)$ and a linear FM chirp (chirping at a much, much slower rate than the transmitted FM chirps) in the form of the term $\pi K\left(\frac{2R(\eta)^2}{c} - \tau_c^2\right)$. This is in contrast to the quadrature demodulated data, which consists of a baseband linear FM in range and a similar slow-chirping linear FM waveform in azimuth, as described in Equation (3.4).

To get a fully formed, fully focused image where each scatterer on the ground is represented by a unique two-dimensional (2D) sinusoid, we will have to do some more processing. This set of steps, typically done on a computer on the ground where computing resources are plentiful, is called *image formation* and will be described at a high-level in the next chapter. Chapter 5 then will examine two specific algorithms for image formation for quadrature demodulated data, the Range-Doppler Algorithm (RDA) and the Omega-K Algorithm (ω-K). Chapter 6 will, in turn, describe Polar Formatting Algorithm (PFA) and Backprojection Algorithm (BPA), which are used for the formation of focused dechirped imagery. It is not difficult to see, given the results of the two examples in

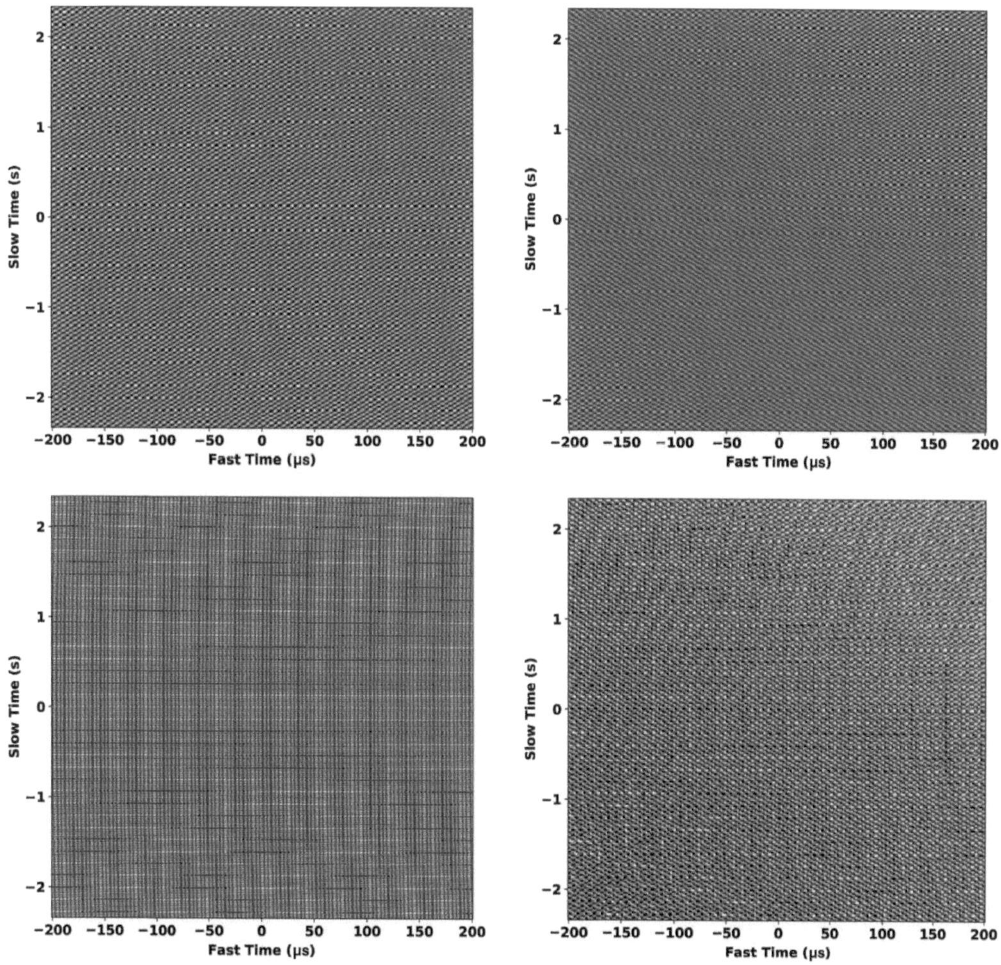

FIGURE 3.10 From top left to bottom right, the plots show the real, imaginary, magnitude, and phase of s_{qd} for the three targets shown in Figure 3.8.

the previous sections, that while the receiver hardware for quadrature demodulation is conceptually simple, it likely requires more processing to get to the fully focused, fully formed image, unlike dechirped data, which already in its sinusoidual form in range and will need a matched filter for the chirp in azimuth along with some interpolation, to form and focus the image.

CHAPTER 4

Image Formation

After the development of the full mathematical model, it becomes clear that more signal processing is needed to acquire the high-resolution radar image. In all cases, the signal processing needed is to correct for the range curvature that is present in the data. The processing chain for this will be presented. Each of the modules of the processing chain will be discussed in some general detail. A brief review of data interpolation will be conducted; interpolation is used extensively in synthetic aperture radar (SAR) image formation algorithms. After the image formation chain is outlined, the first major issue in image formation will be discussed in the next section.

4.1 The Image Formation Chain

There are several different ways to process the raw phase history data after quadrature demodulation or dechirp. Most of the methods take the following steps to complete *image formation processing*. Figure 4.1 shows the different steps needed to form a well-focused, high signal-to-noise ratio (SNR) SAR image, suitable for display.

The first step is to correct for nonlinear, nonconstant velocity motion in the aircraft or spacecraft platform. This correction takes an estimate of the actual motion and applies a series of phase corrections to each pulse based on the actual motion. Next, the phase data is matched filtered, interpolated, and Fourier transformed in several different steps using one of several different image formation algorithms. These algorithms include the Range-Doppler Algorithm (RDA), Omega-K Algorithm (ω-K, sometimes called the "Range Migration Algorithm" [RMA][1]), Polar Formatting Algorithm (PFA), or Backprojection Algorithm (BPA). Typically, RDA and ω-K/RMA are presented as stripmap/quadrature demodulation algorithms,[2] and PFA and BPA as spotlight/dechirp algorithms.[3]

---

[1] See Chapter 12 in W. G. Carrara, R. S. Goodman, and R. M. Majewski, *Spotlight Synthetic Aperture Radar*, Artech House, Boston, MA, 1995.

[2] I. G. Cumming and F. H. Wong, *Digital Processing of Synthetic Aperture Radar Data: Algorithms and Implementation*, Artech House, Boston, MA, 2005.

[3] C. V. Jakowatz, D. E. Wahl, P. H. Eichel, D. C. Ghiglia, and P. A. Thompson, *Spotlight-Mode Synthetic Aperture Radar: A Signal Processing Approach*, Kluwer Academic Publishers, Boston, MA, 1996.

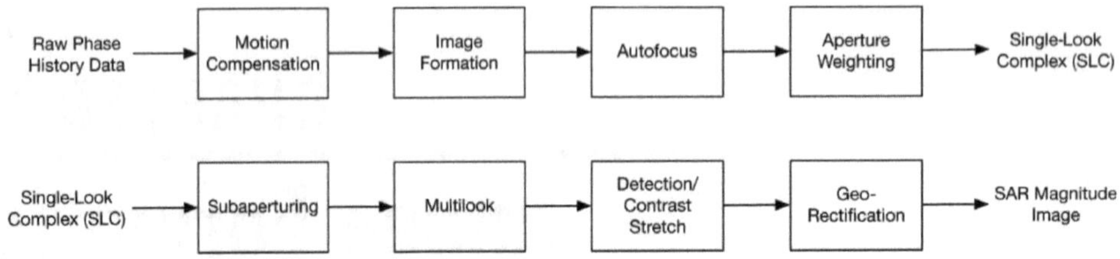

FIGURE 4.1 The flow diagram for image formation processing.

In fact, the reader may be left with the impression that the dechirp techniques, along with Polar Formatting and Backprojection, are only suitable for spotlight mode SAR, while quadrature demodulation techniques, along with algorithms such as Range-Doppler, are only for stripmap. This is not the case. Each of these algorithms, with some minor adaptations, can be used for either stripmap or spotlight and with quadrature demodulation. As was implied in previous chapters, what is being measured over time is the range to a given object at a given location in the scene with respect to the antenna aperture point. Whether the beam remains on a patch of ground continuously (spotlight) or is dragged across the ground (stripmap) only addresses how long the illuminated beam remains on the target. The longer the illumination, the better the resolution, assuming constant linear velocity. Since, in stripmap, the beam is being dragged across the ground, a given object is only in the scene a finite amount of time. In this book, we'll assume that we're only collecting data up to when a target enters and leaves the scene, so as not to get bogged down in stripmap data collection strategies.

While RDA, RMA, PFA, and BPA will be described and discussed in this book, it should not be assumed that these are the only image formation algorithms. For example, another well-known algorithm is the Chirp Scaling Algorithm.[4] Chirp scaling was developed to eliminate the interpolators used for a type of range migration. Until the early 1990s, Range Doppler was the preferred formation algorithm for civilian satellite SAR processing.[5] However, the algorithm suffered from a high computational load when trying to compensate for the range migration using interpolation. Additionally, it was not easy to deal with image formation when the data was taken under high-squint, wide-aperture cases.

For an introduction to SAR, a description of RDA, RMA, PFA, and BPA is sufficient. These are the most common of image formation algorithms. However, each comes with its own advantages and disadvantages. These will be described in Chapters 5 and 6.

After the image has been formed, there may be smearing across the entire aperture in the azimuth direction. This is typically caused by insufficient or inaccurate motion

---

[4] R. K. Raney, H. Runge, R. Bamler, I. G. Cumming, and F. H. Wong, "Precision SAR Processing Using Chirp Scaling," *IEEE Trans. Geoscience and Remote Sensing*, Vol. 32, No. 4, July 1994.

[5] I. G. Cumming and F. H. Wong, *Digital Processing of Synthetic Aperture Radar Data: Algorithms and Implementation*, Artech House, Boston, MA, p. 283, 2005.

compensation. This smearing is a result of an additional phase error being present in the data. This error can be estimated and removed, improving the quality of the overall image. This technique is called *autofocus*. Autofocus will be described in detail in Chapter 7.

Aperture weighting may be optionally applied at this point in the overall algorithm flow. Aperture weighting suppresses or minimizes the effects of raised sidelobes in our return. A single point scatterer, after image formation, takes on the form of a sinc squared function. The largest lobe, which is at the center of the curve, is the *mainlobe*, while the lobes to either side of the mainlobe are the *sidelobes*. If these sidelobes are raised, they can interfere with other mainlobes from other scatterers. Aperture weighting pushes down the sidelobes and provides a sort of sharpening on the image. There are two major techniques to do this, windowing using a function such as a Taylor window and subapodization. Both will be described in a later chapter.

At this point in the process, the data is a fully formed, fully focused SAR image. The resulting product is typically called a single-look complex (SLC) image. This image is suitable for coherent SAR applications or for the creation of a high-quality magnitude image. In the case of moving target indication (MTI) or interferometry, autofocus and aperture weighting are often skipped so as to keep the phase as pristine as possible.

If a high-quality magnitude image is desired, the second set of steps as shown in Figure 4.1 may be performed. The first set is called subaperturing and multilook. These steps create a set of reduced resolution incoherent images from the SLC data. These images are incoherently averaged, boosting each scatterer and pushing down the noise, providing a higher quality SNR. The complex data is then *detected*, which means the magnitude of the complex data is calculated. A contrast stretch is often applied to the data and then an 8-bit grayscale image is produced in a typical image format, such as JPG, TIF, or PNG. This is the detected magnitude product. It represents a visually pleasing representation of the microwave reflectivity of the scene, but it is typically not calibrated nor easily exploitable for things like automatic target recognition (ATR) algorithms.

Finally, the data can be geo-rectified so that it creates an orthonormal image, projected into the ground plane into a suitable Geographical Information Systems (GIS) coordinate system. After image formation, the data lives in the *slant plane*, which is the plane created between the flight track and the projection of the flight track onto the ground at the ground reference point (GRP). The most typical projection is a slant-to-ground plane projection, which puts the data into a flat Earth model, but not necessarily a georeferenced coordinate system. From there, a projection into a suitable coordinate system using an affine transform or the like can be performed, depending on how the data is to be displayed, whether that's in a simple image viewer or a sophisticated GIS tool.

Figure 4.2 shows an orthorectified, multilooked ICEYE-X2 SAR image of the coast of Tanzania. The image has been projected into a WGS-84 coordinate system to provide latitude and longitude coordinates for each pixel in the scene. The image has exquisite oceanographic features, which are highlighted after the additional processing steps of multilook and contrast stretching.

FIGURE 4.2 An orthorectified, multilook ICEYE-X2 image near Cordoba, Argentina. The polarization is VV. This image went through the entire image processing chain to create a visually pleasing, human exploitable synthetic aperture radar image product.

4.2 Motion Compensation

In developing the mathematical model for SAR, we have assumed that the aircraft or spacecraft flies straight and level with constant velocity v_a. Spacecraft do not fly straight and level, although the velocity is often very constant. Depending on the total time of the dwell, it may be acceptable to assume that the spacecraft's flight is more or less straight over the small section of the total orbit.

Aircraft, on the other hand, are not nearly so well behaved. Depending on the flight geometry and imaging mode desired, the craft is almost certainly not going to be flying straight and level at a constant velocity. When this occurs, the mathematical model upon which the SAR processing is built breaks down. The image quality is likely to degrade. If the deviations from a straight and level path with a constant velocity are relatively large, the image will likely be so deformed as to be unrecognizable. It is possible to correct for these effects using two different techniques: motion compensation and autofocus.

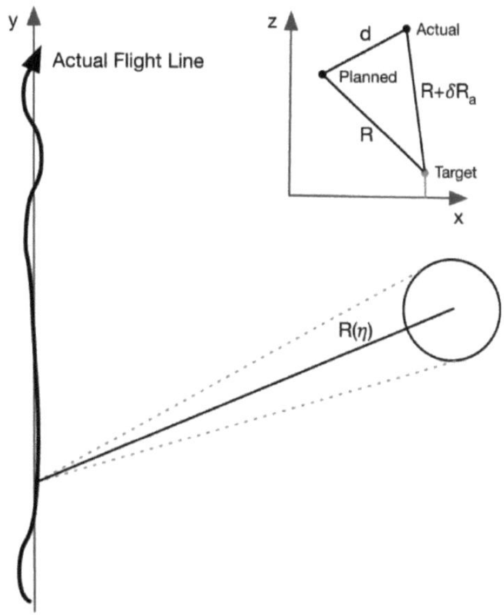

FIGURE 4.3 Diagram showing the geometrical model for nonlinear flight paths. The drawing on the left is a top-looking view where the *y*-axis shows the intended straight and level flight path, while the dark black line shows the actual flight path. The inset diagram in the upper right is a side-looking view where the flight path is going into the page. The black dot on the left is the planned position of the sensor on the straight, level flight path. The black dot on the right is the actual position.

Motion compensation[6] takes the kinematic information from the aircraft and corrects the differences in range that occur from a non-linear flight path. These corrections help maintain coherency across the aperture. Without the correction, phase errors can occur in the model. These phase errors are generally quadratic in nature and will cause smearing across the aperture.

Errors in velocity can be corrected using autofocus techniques.[7] This will be examined in more detail in a later chapter.

Figure 4.3 illustrates the geometrical model for nonlinear flight paths. The drawing on the left of the figure is a top-looking view where the *y*-axis shows the intended straight, level flight path. The thick line in black shows the actual nonlinear flight path. The inset diagram in the upper right of the figure is a side-looking view where the flight path is going into the page. The inset diagram represents the planned versus actual position of the aircraft for each pulse during imaging. The black dot on the left of that inset is the planned position of the sensor on the straight, level flight path. The black dot on the right is the actual position at that time. The result is a range error of δR_a.

---

[6] G. Fornaro, G. Franceschetti, and S. Perna, "Motion Compensation Errors: Effects on the Accuracy of Airborne SAR Images," *IEEE Trans. Aerospace Elec Systems*, Vol. 41, No. 4, October 2005.

[7] D. E. Wahl, P. H. Eichel, D. C. Ghiglia, and C. V. J. Jakowatz, "Phase Gradient Autofocus-a Robust Tool for High Resolution SAR Phase Correction," *IEEE Trans. Aerospace Elec Systems*, Vol. 30, No. 3, July 1994.

The motion and position of the aircraft is measured by an Inertial Navigation System (INS), which likely consists of a Global Positioning System (GPS) unit and an Inertia Measurement Unit (IMU). The IMU generally consists of a set of three-axis gyros that provide the kinematics (velocity, acceleration, roll, pitch, and yaw), while the GPS unit provides the coordinates in a three-dimensional space, typically in a latitude, longitude, and altitude coordinate space, although the GPS unit may be able to measure and report Earth-Centered-Earth Fixed (ECEF) coordinates, which are in meters with respect to the center of Earth, making range error corrections easier and potentially more accurate. The INS is a system that incorporates all of kinematic and position measurements and provides estimates of future motion based on past motion using something like a Kalman filter.

Once an accurate estimate of the position and motion of the aircraft is achieved, the procedure to compensate for a nonlinear flight path is straight forward. The δR_a is calculated for each pulse in the dwell and a phase correction of

$$M(\eta) = \exp\left(j\frac{4\pi}{\lambda}\delta R_a(\eta)\right) \tag{4.1}$$

where λ is the center wavelength of the radar system and $\delta R_a(\eta)$ is the range offset at some time η. This correction is generally applied after the range compression stage, which we'll describe in more detail in the coming chapters. A second correction may be applied in later parts of the processing, generally during the step where the curvature of range is corrected. This curvature, called *range migration*, occurs when a target drifts from one range cell to another across the aperture. A range cell is one specific range distance from the sensor to a spot on the ground. A pulse consists of N_r range cells, where N_r is the number of range samples taken in a pulse.

4.3 Processing Tools

Synthetic aperture radar processing requires several signal processing tools. Two of the most ubiquitous of tools are the Fourier transform and interpolation. A brief review of both along with practical tips on how to implement those tools in MATLAB and Python will be discussed in this section. This section, and indeed this book, assumes the reader has experience with digital signal processing.

4.3.1 Fourier Transforms

Fourier transforms are arguably the most important tools in the digital signal processing toolbox for SAR processing. The transform takes a time series and identifies its frequency components, providing their magnitude and phase. Since we are tracking individual scatterers in a SAR scene by their differing frequencies from each other, the Fourier transform seems like a natural tool to sift through the superposition of sinusoids. The continuous-time Fourier transform is defined as

$$S(f) = \int_{-\infty}^{+\infty} s(t)e^{-j2\pi ft}dt \tag{4.2}$$

where j is the designator for an imaginary number, where $j = \sqrt{-1}$. The inverse Fourier transform is thus defined as

$$s(t) = \int_{-\infty}^{+\infty} S(f)e^{+j2\pi ft}df \tag{4.3}$$

Consider a generic signal $s(t)$. In a radar receiver, this signal is analog and continuous until it reaches the analog-to-digital converter. This converter samples the signal at regular intervals to get a digital number representation of the signal. This signal is typically represented by $s[n]$, where n is the digital index for each sample. The fidelity of the signal is governed by the Nyquist sampling theorem.[8] The theorem states that, for a real-valued bandlimited signal, $s(t)$ is uniquely represented by its samples $s[n] = s(nT)$, $k = 0, \pm1, \pm2, \ldots$ if the sampling frequency $f_s = \frac{1}{T}$, when $f_s > 2f_{\max}$, where $f_{\max}$ is the maximum frequency in the bandlimited signal. For a complex-valued bandlimited signal, $f_s > f_{\max}$ to unique represent $s(t)$.

The Discrete Fourier transform (DFT) is the dual of the continuous-time Fourier transform. It decomposes a finite sequence of equally spaced samples of a function into its constitute frequency components. As we will see in Chapter 5, the "equally spaced samples" will be important. The mathematical definition of the DFT is

$$G[k] = \sum_{n=0}^{N-1} g[n]e^{-j\frac{2\pi kn}{N}} \tag{4.4}$$

and the inverse DFT is defined as

$$g[n] = \sum_{k=0}^{N-1} G[k]e^{+j\frac{2\pi kn}{N}} \tag{4.5}$$

It's clear that calculating these sequences explicitly requires an iterative loop for each value of k (for the forward transform) or n (for the inverse transform), making this a computationally expensive calculation. To get around this, the Fast Fourier Transform (FFT) was developed.

The FFT is an algorithm that computes the DFT of a given sequence (or its inverse) using a series of factorizations. The FFT reduces the complexity from $O(N^2)$ to $O(N \log N)$. There are several ways to do the factorization. In most cases, the FFT calculation accurately represents the DFT, but it may suffer from finite-precision floating point arithmetic errors.

Example 4.1 Consider a signal with two distinct sinusoids, defined by

$$s(t) = 2\cos\left(2\pi f_1 t\right) + \cos\left(2\pi f_2 t\right) \tag{4.6}$$

where $f_1 = 100$ Hz and $f_2 = 200$ Hz. Since this is a real-valued function, the sampling frequency must be greater than twice the highest frequency, which is 200 Hz. Let's consider $f_s = 1400$ Hz, which is seven times the highest frequency. Thus, Figure 4.4 shows $s[n]$ from $t = [0, 0.01]$ seconds.

---

[8] A. V. Oppenheim and R. W. Schafer, *Discrete-Time Signal Processing*, Prentice Hall, Upper Saddle River, NJ, 1989.

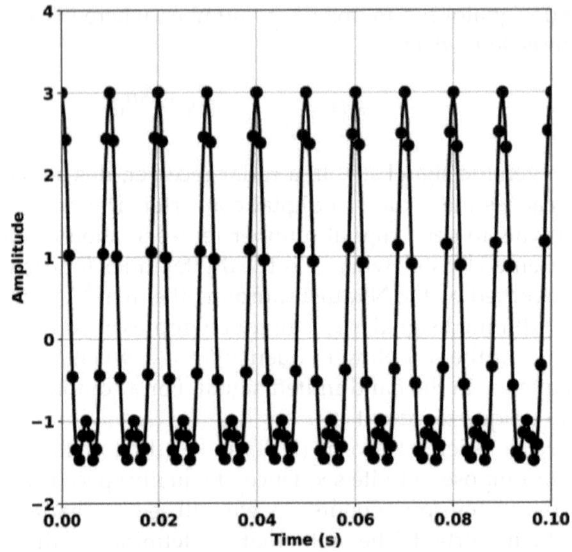

FIGURE 4.4 Properly sampled double sinuodial signal at 7x the highest frequency. The solid line represents s(*t*), while the dots represent the discretely sampled s[*n*].

The solid line in Figure 4.4 represents $s(t)$, while the discrete dots represent $s[n]$. It's clear that at seven times the highest frequency, the waveform is well-represented and easily recognizable.

The frequency response of $s[n]$, $S[k]$, can be easily calculated using the FFT. The basic code for this in MATLAB is

```
f = linspace(-f_s / 2, f_s / 2, N)
S = fftshift(fft(s))
```

where N is the total number of samples, f_s is the sampling frequency, and s is the time series signal. Thus, S is the frequency response of that time series. The vector f represents the frequency span of S.

In contrast, the basic code for this in Python is

```
f = numpy.fft.fftshift(numpy.fft.fftfreq(N, d = 1/f_s))
S = numpy.fft.fftshift(numpy.fft.fft(s))
```

Note that the FFT routines for Python are in the Numpy package. It may be easier to establish a set of anonymous functions to do the FFT, rather than having to type `numpy.fft.` before each command. Or one can include the function from the package directly, for example:

```
from numpy.fft import fft, fftshift, fftfreq
```

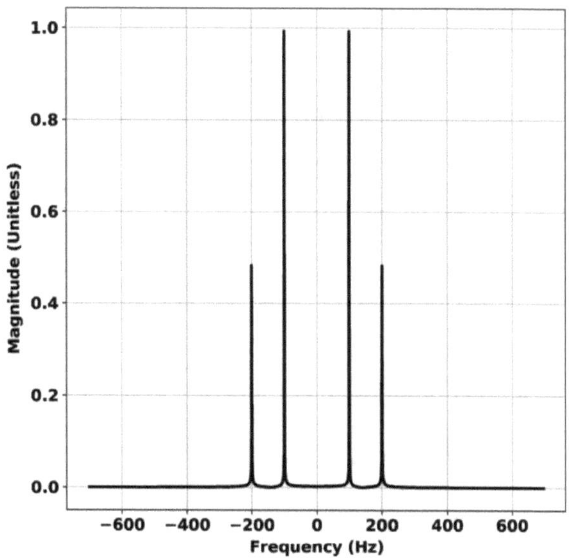

Figure 4.5 The magnitude of the frequency response $S[n]$. Note that there are peaks at $\pm f_1$ and $\pm f_2$. The double peaks represent the time series function's real value nature. The magnitude of the peaks add up to the total magnitude at each frequency pair.

The `fftshift` is a command that puts the DC frequency component at the center of the vector. The FFT function typically calculates the frequencies so that the first half of the sequence are the positive frequencies and the second half are the negative frequencies (but not mirrored). One half of the spectrum, typically the positive half, is called a *one-sided Fourier transform*. The `fftshift` provides a way to restore the sequence to its proper order.

The magnitude $|S[n]|$ is shown in Figure 4.5. This plot shows peaks at $\pm f_1$ and $\pm f_2$. The double peaks represent the time series function's real value nature. Since there is no way to know where, in phase, you are on the sinusoid in a real-valued sinusoid, this results in "negative" frequencies in the frequency response. The magnitude of the FFT needs to be scaled by N to be properly represented.

Example 4.2 Consider the same sinsoidual signal as was presented in the previous example. Instead of sampling at $f_s = 7f_2$, consider an improperly sampled signal at $f_s = 1.3f_2$, which violates the Nyquist sampling theorem.

Figure 4.6 shows $s(t)$ as a solid line, while the improperly sampled $s[n]$ is represented by discrete dots. It's clear that not all of the shape of the continuous waveform is represented in the discrete sampling. Figure 4.7 shows the DFT for $s[n]$. There is a peak at 100 Hz and a peak just above 50 Hz. There is no peak at 200 Hz because there is no sampling frequency that goes to 200 Hz. Instead, the DFT/FFT is displaying a phenomenon called *aliasing*, where the frequency information isn't lost in the transform, but is instead wrapped around on the frequency axis. That is, the decomposition of the higher frequency components is improperly attributed to lower

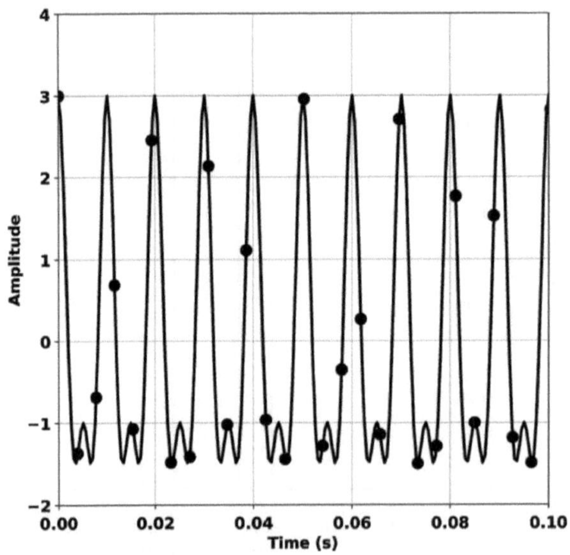

Figure 4.6 The improperly sampled s(t) at $f_s = 1.3f_2$.

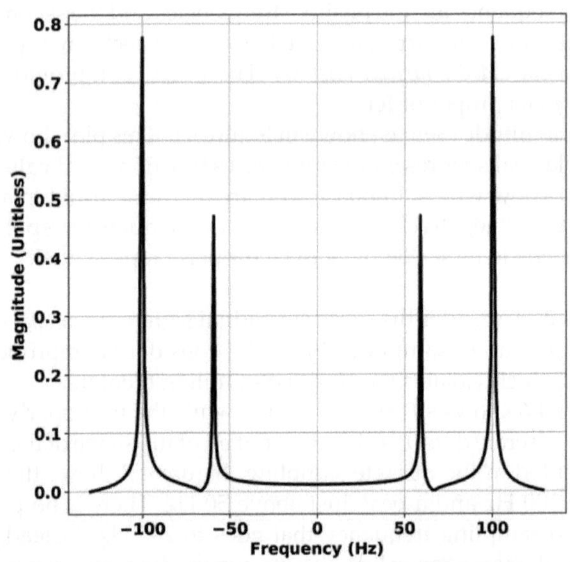

Figure 4.7 The FFT of s[n].

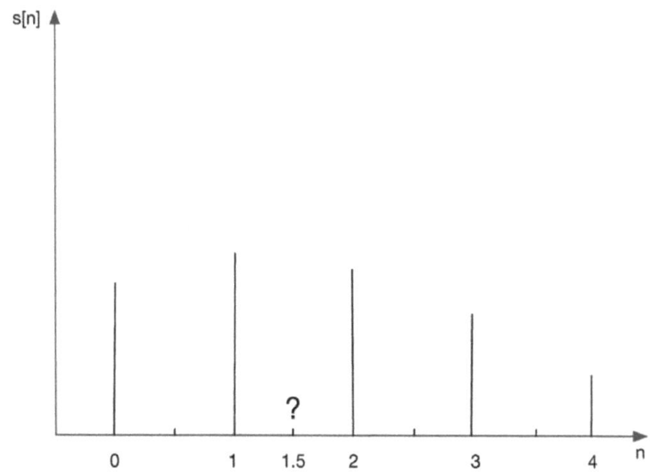

FIGURE 4.8 A generic waveform $s[n]$. If the samples are spaced at $n = [0, 1, 2, 3, 4]$, what should the value at $n = 1.5$ be?

frequencies, giving the impression that there are lower frequency components in the signal when there are none.

4.3.2 Interpolation

Interpolation is the act of calculating new data points typically between two distinct points in a discrete sequence. That is, interpolation fills in the gaps between data points in a discretely sampled function. Figure 4.8 shows a generic discrete signal, $s[n]$. The values of $s[n]$ were sampled at $n = [0, 1, 2, 3, 4]$.

If the value at $n = 1.5$ is desired, then we must turn to interpolation to estimate what this value should be if we had the opportunity to sample it at that instance in time. There are several methods of interpolation, which include linear, piecewise-constant, cubic, nearest neighbor, and polynomial interpolation, just to name a few. Both MATLAB and Python have built-in interpolation functions. In recent years, these interpolation functions have changed in both form and syntax for both MATLAB and Python, so it pays to read the function's documentation page carefully before using it.

Example 4.3 Consider the finely sampled function $s(t) = 2\cos(2\pi f_1 t) + \cos(2\pi f_2 t)$ from the previous section, where the sampling frequency f_s was set to seven times the highest frequency, or, $f_s = 7f_2 = 1400$ Hz. Consider the possibility that a higher sampling frequency is required. One way to do this is to use interpolation. Figure 4.9 shows the results by taking the time sequence $n = [0, 0.0007, 0.0014, \ldots]$ and interpolating to a finer grid $n = [0, 0.00006, 0.000125, \ldots]$. The gray dots are the original discrete sequence with an offset of $+1$ to distinguish it from the newly sampled

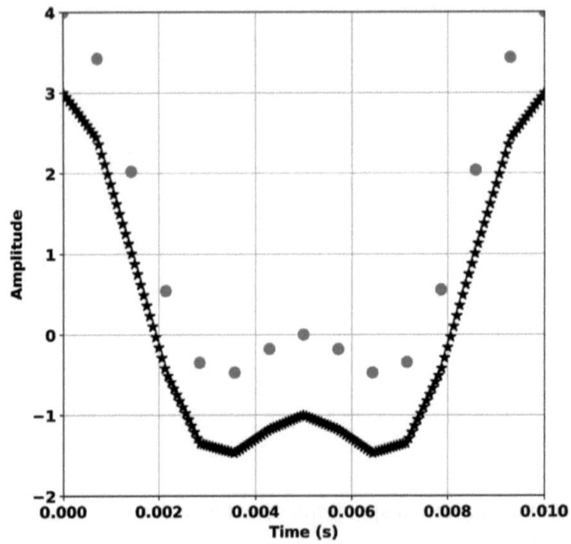

FIGURE 4.9 Interpolation of s[k] from $f_s = 1400$ Hz to $f_s = 1600$ Hz. The gray dots indicate the original s[k] with an offset of +1, while the black stars indicate the interpolated signal.

waveform, shown in black stars in Figure 4.9. The `interp1d` function in the Python package Scipy was used (there is a similar function in MATLAB). A linear interpolation was used to generate the new function values.

CHAPTER 5

Image Formation Algorithms for Quadrature Demodulated Data

In the previous chapters, we conceptually and mathematically discussed how the radar signal propagates to the ground, across the targets, and back to the sensor. In this chapter, we will again examine the quadrature demodulated data. For simplicity, motion compensation will not be specifically addressed here. Two specific image formation algorithms will be presented: Range-Doppler Algorithm (RDA) and Omega-K (ω-K), also known as the Range Migration Algorithm (RMA). The application of these algorithms upon quadrature demodulated data will produce focused synthetic aperture radar (SAR) data, assuming perfect flight geometries.

5.1 Range-Doppler Algorithm

The Range-Doppler Algorithm has a rich history of processing SAR data. It was first developed at the Jet Propulsion Laboratory in 1976 by Wu.[1] It was used to focus the first spaceborne SAR image in 1978[2] from the SEASAT SAR system. SEASAT was launched June 27, 1978 and its mission ended on 10 October of the same year, after a massive short circuit in the electrical bus system.[3] The algorithm still remains in wide use today, owing to its relatively simple algorithmic form under low-squint imaging. In 1984, Jet Propulsion Laboratories modified RDA to handle moderate amounts of

[1] C. Wu, "A Digital System to Produce Imagery from SAR Data," *AIAA Conference: System Design Driven by Sensors*, October 1976.

[2] J. R. Bennett, I. G. Cumming, R. A. Deane, P. Widmer, R. Fielding, and P. McConnell, "SEASAT Imagery Shows St. Lawrence," *Aviation Week and Space Technology*, 26 February 1979.

[3] "Report of the SEASAT Failure Review Board," *Readings in Systems Engineering*, NASA Scientific and Technical Information Program, NASA-SP-6102, 1993.

squint.[4] The moderate amounts of squint produce large range migration. To effectively compensate for this effect, the secondary range correction (SRC) routine was added to the algorithm. The SRC compensates for the coupling of range and azimuth in the phase history; it is applied after range compression.

5.1.1 Algorithm Overview

Figure 5.1 describes the algorithmic steps necessary to complete the RDA. This flow is for low-squint imaging cases, where the squint is low enough that the range equation can safely be assumed to be parabolic with time. If this assumption is violated, the more accurate forms of RDA with SRC may be needed. We will consider the low-squint case here.

The algorithm starts with range compression. In this step, an fast Fourier transform (FFT) is performed along each received pulse ($s(t,k)$, where k represents an indexing for each slow time pulse), followed by a range matched filter, then a subsequent inverse fast Fourier transform (IFFT). The steps remove the baseband linear FM chirp for each scatterer in the scene. The output of the matched filter, which is done in frequency space so that a simple element-by-element vector multiply can be performed, is a sync where each linear FM chirp is detected.

The next step is to take an FFT along each range bin across azimuth ($s(k,\eta)$, where k represents an indexing across each range bin). Once this step is complete, the data is said to be in the *range Doppler* domain. It is in this domain that Range Cell Migration Compensation (RCMC) is performed. RCMC is range time/azimuth frequency dependent, straightening out target trajectories at the same range so that they run parallel to the Doppler axis.

Once RCMC has been applied, then a matched filter in azimuth can be applied. This compresses the azimuth linear FM chirp. The matched filter can be done again with an element-by-element multiply across azimuth at each range bin. Finally, an IFFT puts the data back into range-azimuth space, resulting in a single-look complex image. This image is sometimes called "compressed" because of the two matched filter steps. This is not to be confused with image or data compression where the amount of data is reduced to reproduce either exactly or approximately some data set or image. The compressed single-look complex image can now be detected by taking the magnitude of the complex data, producing an uncalibrated microwave reflectivity map of the scene. That is, the image that we've been seeking.

5.1.2 Raw Radar Data

Recall from previous chapters that our data, after it has been demodulated to baseband, follows the form

$$s_{\text{qd}}(t,\eta) = Aw_a[t-\tau(\eta)]\,\text{rect}\left(\frac{t-\tau(\eta)}{T}\right)e^{j\pi(4f_0\tau(\eta)-K(t-\tau(\eta))^2)} \tag{5.1}$$

where A is the attenuation based on radar reflectivity and attenuation through the atmosphere and is complex, w_a is the antenna beam pattern, $\tau(\eta)$ is the change in round-trip

[4] M. J. Lin and C. Wu, "A SAR Correlation Algorithm with Accommodates Large Range Migration," *IEEE Trans Geoscience and Remote Sensing*, Vol. 22, No. 6, November 1984

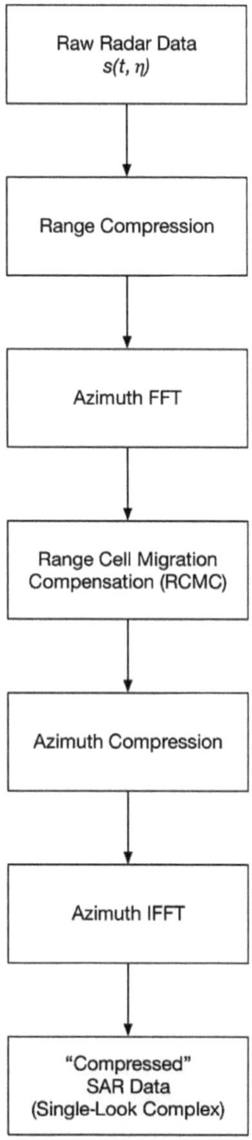

FIGURE 5.1 The algorithm flowchart for the Range-Doppler Algorithm (RDA) under low-squint conditions.

time delay from an arbitrary scatterer to the sensor over slow time, K is the linear FM chirp rate, f_0 is the center frequency of the radar, and t and η are fast and slow times, respectively. The rectangular function is the range envelope that defines the time period in which the receive signals from the scene are captured. This equation can be rewritten in terms of range to the target

$$s_{\text{qd}}(t,\eta) = Aw_a[t - (2R(\eta)/c)]\,\text{rect}\left(\frac{t - (2R(\eta)/c)}{T}\right) \cdot e^{j\pi(4f_0(2R(\eta)/c) - K(t-(2R(\eta)/c))^2)} \tag{5.2}$$

Parameter Name	Variable	Value/Units
Slant range to GRP	$R(\eta_c)$	20 km
Radar velocity	V_a	150 m/s
Pulse duration	T	2.5 μs
Range chirp rate	K	20 MHz/μs
Center frequency	f_0	5.3 GHz
Doppler bandwidth	Δf_{dop}	80 Hz
Range sampling rate	F_r	60 MHz
Azimuth sampling rate	F_a	100 Hz
Beam squint angle	θ	0°
Beam center crossing time	η_c	0 s
Number of range lines	N_a	256 lines
Number of samples per range line	N_r	320 samples

Table 5.1 Parameters for C-Band Airborne Synthetic Aperture Radar

We'll step through each part of the RDA using an example. Four scatterers will be placed in 400-meter by 400-meter flat surface, with a SAR flying at a slant range of 20 kilometers away from the scene center. Table 5.1[5] shows the parameters for the simulated airborne radar. The four scatterers are placed in such a way that one is at the center of the scene (at the GRP), two are in line with each other in range, and two are in line with each other in azimuth. Figure 5.2 shows the placement and the simulated antenna beam pattern at the center of the dwell, at $\eta_c = 0$ second.

Using the algorithmic steps as described in Chapter 3, the simulated raw data, along with the antenna beam pattern and the round-trip delay time for each pulse, is shown in Figure 5.3. These are the results of the simulation with the four scatter points as described in Figure 5.2. The plot on the top left illustrates the antenna beam pattern across slow time for each of the targets; the top right plot shows the pulse width and round-trip time delay for each of the scatterers. It is worth noting that two of the target's pulses overlap; these are the targets with the same range position. The plot on the bottom left shows the magnitude of the resulting simulated phase history, whereas the bottom right shows the phase of the simulated data. If you look at the plot on the bottom left carefully, it is clear that there is an "X" shape for each of the targets in relative placement to each other. The X shape is due to an up chirp. If the chirp rate was negative and the chirp was a downward ramp, these "X" shapes would be more like a bullseye.

[5] Inspired by Table 6.1 in I. G. Cumming and F. H. Wong, *Digital Processing of Synthetic Aperture Radar: Algorithms and Implementation*, Artech House, p. 228, 2005.

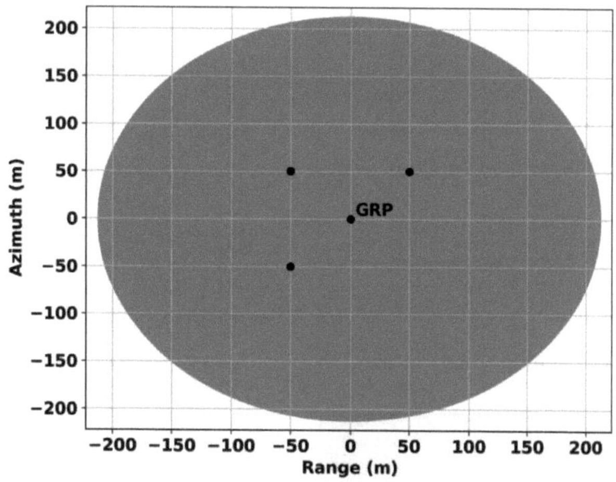

FIGURE 5.2 Placement of four scatterers in simulated ground plane. One scatterer is placed at ground reference point (GRP), while the others are placed in such a way that two are in line in range and two are in line in azimuth.

5.1.3 Step 1: Range Compression

The first step in the RDA is to perform range compression. This is accomplished by taking an FFT at each pulse, multiplying that by a matched filter, and then transforming the data back into range with an IFFT. The matched filter could be done directly in the range dimension, rather than in the range frequency dimension, by implementing the matched filter as a convolution rather than as a multiplication in frequency space. Since the FFT-multiplication-IFFT is computationally and conceptionally efficient, we'll describe that process here.

When the FFT is taken, it can be zero padded with the padding placed at the end of the replica array. This is typically done so that the FFT size is a power of 2, which improves computational performance in some FFT algorithms. The FFT is performed on each η_k pulse, where $k = 0, \ldots, N_a$, where each pulse is of the form of Equation (5.2). The transformed data is represented by

$$S_0(f_t, \eta_k) = FFT|_{k=[0,N_a-1]}\{s_{\mathrm{qd}}(t, \eta_k)\} \tag{5.3}$$

where f_t is range frequency. This is a one-dimensional FFT along each pulse. The matched filter is defined as

$$G(f_t) = \mathrm{rect}\left(\frac{f_t}{|K|T}\right) e^{-j\pi\frac{f_t^2}{K}}, \tag{5.4}$$

where K is the chirp rate and T is the pulse width. The width of the rectangular window is defined by the radar bandwidth ($B = |K|T$). Range frequency is defined by $f_t = Kt$ and, with some rearranging, $t = \frac{f_t}{K}$. Figure 5.4 shows the real component of the range matched filter.

The matched filter is completed by multiplying the filter G with S_0.

$$S_{\mathrm{rc}}(f_t, \eta_k)|^{k=[0,N_a-1]} = G(f_t)S_0(f_t, \eta_k) \tag{5.5}$$

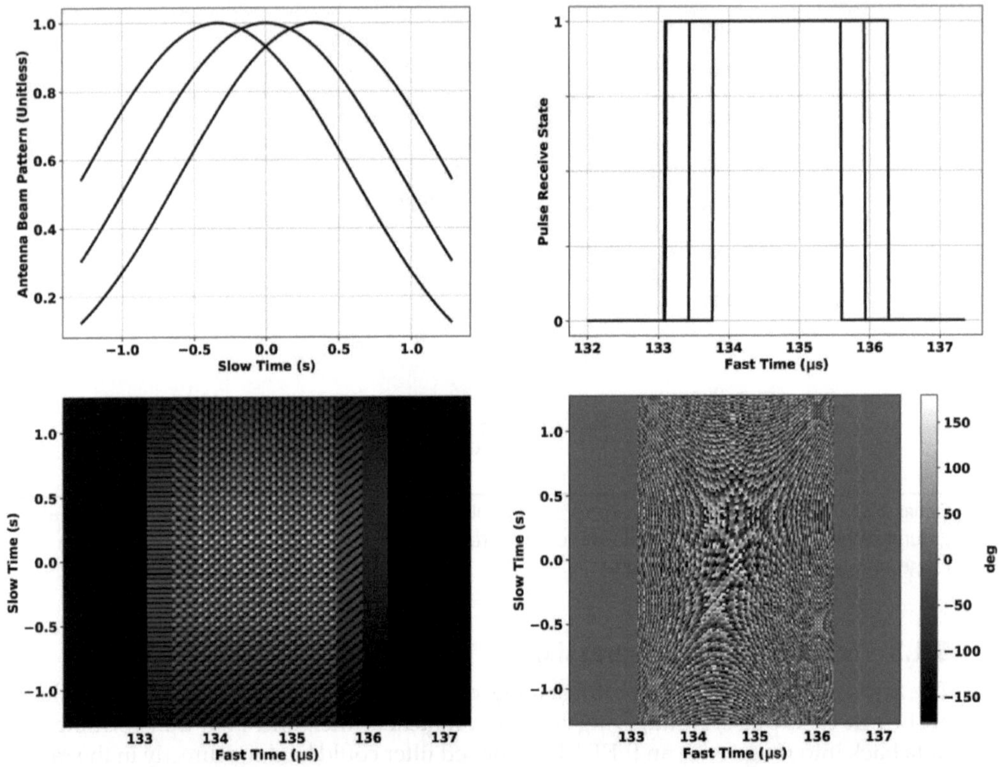

FIGURE 5.3 The simulated data from the four scatterers. In the top left, the antenna beam pattern across slow time for each of the scatterers. Top right shows the pulse width and round-trip time delay for each of the scatterers; note that two of the target's (the targets with the same range position) pulses overlap. The bottom left shows the magnitude of the resulting simulated raw data; the bottom right shows the phase of the resulting simulated raw data.

The range compression is completed by taking a one-dimensional inverse Fourier transform along each pulse to put the data back into range.

$$s_{\mathrm{rc}}(t,\eta) = \mathrm{IFFT}|^{k=[0,N_a-1]}\{S_rc(f_t,\eta_k)\} \tag{5.6}$$

The resulting output of this step is a sync-like envelope in the range direction. This is described by

$$s_{\mathrm{rc}}(t,\eta) = A_0 p_r\left[t - 2R(\eta)/c\right] w_a\left(\eta - \eta_c\right) e^{-j4\pi f_0 R(\eta)/c} \tag{5.7}$$

where A_0 is the overall gain of a specific scatterer in the scene, w_a is the term for the antenna beam pattern, p_r is the resulting sync-like envelope, which is varying over slow time, and the exponential term is the slowly varying phase over slow time. The sync-like function and phase vary as range to a specific scatterer changes. For example, s_{rc} is actually a superposition of these sync plus phase functions for all four scatterers. Figure 5.5 shows the real component and the magnitude for s_{rc}, which is our radar signal after

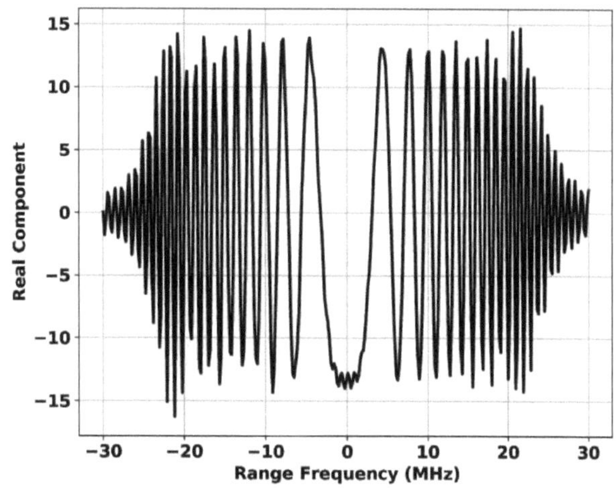

FIGURE 5.4 Real component of the range matched filter $G(f_t)$.

range compression. Looking carefully at the plot on the left, the sync signal can be seen in the light-dark patterns along each range bin for each target. The targets on the left are brighter because there are two targets on the same range line that essentially doubles the energy in that range bin. The rest of the steps in the RDA are designed to focus the data in the azimuth direction by compensating for the range varying terms in Equation (5.7).

5.1.4 Step 2: Azimuth FFT

The next step in the algorithm is to take an FFT along each range bin in $s_{rc}(t,\eta)$,

$$S_1(t,f_\eta) = \mathrm{FFT}|_{j=[0,N_r-1]}\{s_{rc}(t_j,\eta)\} \tag{5.8}$$

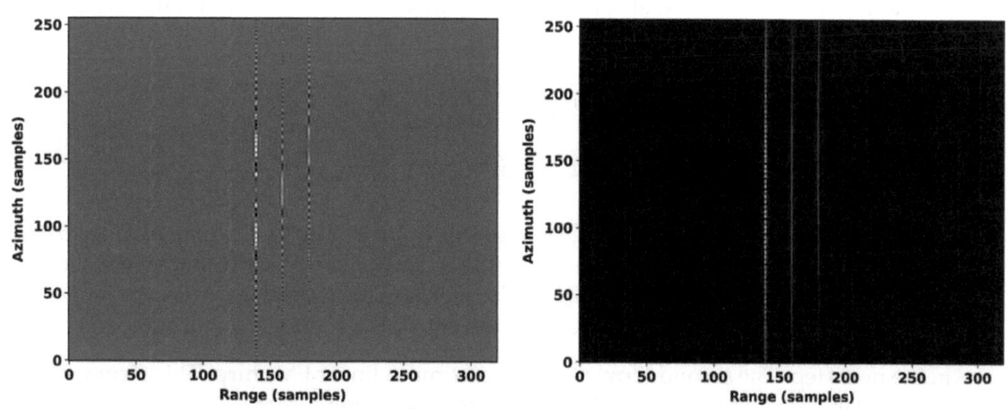

FIGURE 5.5 Resulting signal after range compression, with the real component on the left and the magnitude on the right.

To understand the form that $S_1(t,f_\eta)$ takes, we need to examine Equation (5.7). For low-squint cases in stripmap imaging mode, the range equation can be approximated by a parabolic equation, provided that η isn't too large (i.e., the synthetic aperture isn't too large). This approximation takes the form

$$R(\eta) = \sqrt{R_0^2 + v_r^2\eta^2} \approx R_0 + \frac{v_r^2\eta^2}{2R_0} \tag{5.9}$$

where R_0 is the slant range of closest approach of that given target or location.

From this approximation, we can write Equation (5.7) as

$$\begin{aligned} s_{rc}(t,\eta) &\approx A_0 p_r\left[t - 2\frac{2R(\eta)}{c}\right] w_a(\eta - \eta_c) \\ &\cdot e^{-j\frac{4\pi f_0 R_0}{c}} e^{-j\pi\frac{2v_r^2}{\lambda R_0}\eta^2} \end{aligned} \tag{5.10}$$

where λ is the center wavelength of the radar. The slowly varying range is now plain, showing itself in the second exponential term $e^{-j\pi\frac{2v_r^2}{\lambda R_0}\eta^2}$. This implies that the azimuth chip rate can be calculated by

$$K_a \approx \frac{2v_r^2}{\lambda R_0} \tag{5.11}$$

It's clear that the second term is the critical one for each scatterer, as the first term is a constant phase term over the dwell. As far as the second exponential term goes, Step 4 in the RDA is designed to compress this chirp into a single sync-like function, similar to what occurred in Step 1.

The azimuth FFT in Step 2 transforms the data into what is commonly referred to as the *range Doppler domain*. Using the Principle of Stationary Phase (POSP)[6], the relationship between azimuth frequency and slow time is

$$f_\eta = -K_a\eta \tag{5.12}$$

Taking $\eta = -f_\eta/K_a$ and substituting into Equation (5.8), the radar data after Step 2 can be represented by

$$\begin{aligned} S_1(t,f_n) &= A_0 p_r\left[t - \frac{2R_{rd}(f_\eta)}{c}\right] W_a(f_\eta - f_e ta_c) \\ &\cdot e^{-j\frac{4\pi f_0 R_0}{c}} e^{j\pi\frac{f_\eta^2}{K_a}} \end{aligned} \tag{5.13}$$

The two exponential terms are important in this domain. The first exponential term carries the coherent information on each scatterer and is useful for applications such as interferometry, which we will discuss in Chapter 11. This term also carries the range cell migration (RCM) R_{rd}. It is this range–azimuth coupled term that we will compensate for in the next step. The second term $e^{j\pi\frac{f_\eta^2}{K_a}}$ is the azimuth linear FM chirp. This term will get removed in Step 4, leaving only the phase information per target.

[6]A. Papoulis, *Signal Analysis*, McGraw-Hill, New York, NY, 1977.

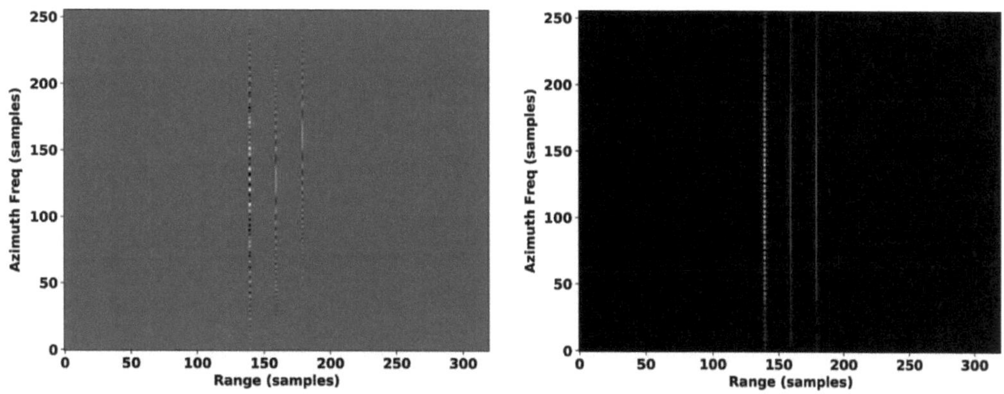

FIGURE 5.6 The data in the range-Doppler domain, after the azimuth FFT. The real component of S_1 is on the left, while the magnitude is on the right.

Figure 5.6 shows the radar data after the azimuth FFT; the data resides in the range-Doppler space after this step. Note that the two targets on the same range line are now interfering with each other, which is clear from the modulation in the line of targets on the left of the magnitude plot. Such a modulation is not present in the middle or right lines, which contain the energy from the scatterer at the GRP and the farthest scatterer, respectively.

5.1.5 Step 3: Range Cell Migration Compensation

Equation (5.13) contains the term $R_{\mathrm{rd}}(f_\eta)$. This term describes the change in range with respect to the scene center over slow time (or, in this case, azimuth frequency). Combining f_η with the parabolic approximation of range R_{rd} can be approximated by

$$R_{\mathrm{rd}}(f_\eta) \approx R_0 + \frac{\lambda^2 R_0 f_\eta^2}{8v_r^2} \tag{5.14}$$

The second term in this relationship is the change in range over azimuth frequency, which we will call ΔR,

$$\Delta R(f_\eta) = \frac{\lambda^2 R_0 f_\eta^2}{8v_r^2} \tag{5.15}$$

For our four scatterers case, the change in range is shown in Figure 5.7.

There are several different ways to implement the RCMC. One of the easiest involves interpolating each pulse to correct for the range curvature. This can be done by converting ΔR into approximate range bins: $b \approx \Delta R/\rho_a$, where ρ_a is the slant plane resolution in azimuth. The variable b represents the number of range bins, or range cells, to shift in range. Figure 5.8 shows the number of bins to shift for each pulse. A simple 1D

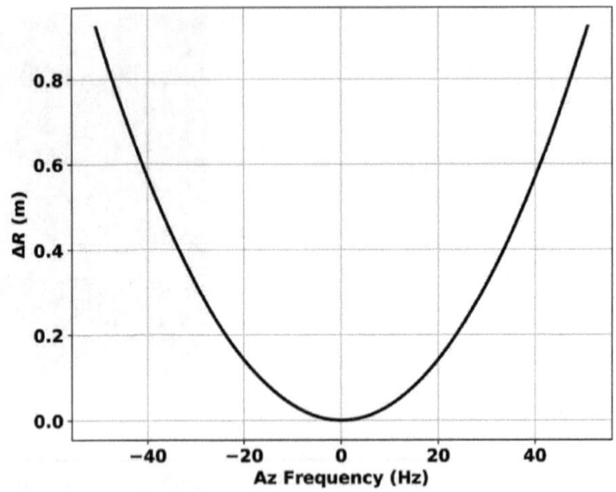

FIGURE 5.7 ΔR versus azimuth frequency for the example with four scatterers.

interpolation can be performed for each pulse to perform the shift. After this is done, the range Doppler data should look like Figure 5.9 and is represented by

$$S_2(t, f_\eta) = A_0 p_r \left(t - \frac{2R_0}{c}\right) W_a \left(f_\eta - f_{\eta_c}\right) \cdot e^{-j\frac{4\pi f_0 R_0}{c}} e^{j\pi \frac{f_\eta^2}{c}} \tag{5.16}$$

The absence of the R_{rd} term in p_r indicates that the compensation corrected the RCM. If the term had not be corrected, then the range varying sync-like envelope would be

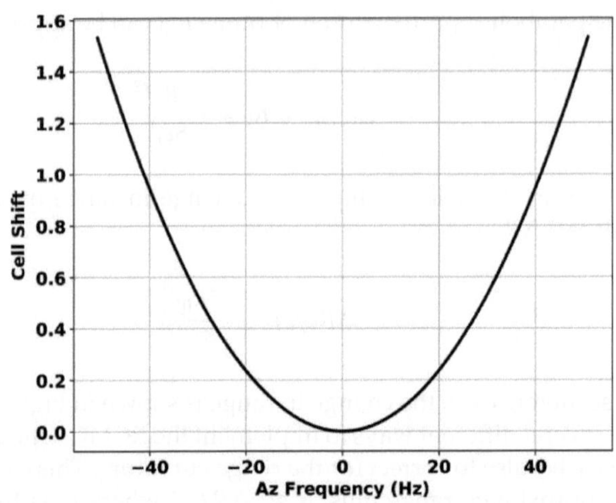

FIGURE 5.8 The cell shift b for each pulse across the aperture.

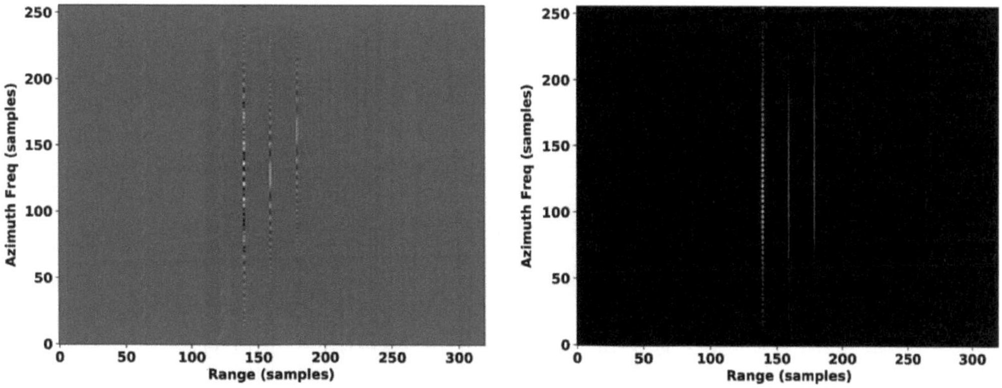

FIGURE 5.9 The data $S_2(t, f_\eta)$ after range cell migration compensation (RCMC). The image on the left is the real component of S_2, while the magnitude image of S_2 is on the right.

distorted, smearing in the azimuth direction after the rest of the algorithm had been completed. While the RCM in this particular example is slight (it is difficult to see the curvature in Figure 5.6), it is still present in the data and would have affected the quality of our final image if not corrected. Figure 5.10 shows the difference between the real components of S_1 and S_2. The plot, zoomed into the area of the target energy for visual interpretation, shows the range curvature. It's clearly corrected in S_2, after RCMC is performed.

5.1.6 Step 4: Azimuth Compression

A matched filter is applied to S_2 after RCMC is performed to compress the linear FM chirp $e^{j\pi \frac{f_\eta^2}{c}}$. This matched filter takes on the form

$$H_{\text{az}}(f_\eta) = e^{-j\pi \frac{f_{eta}^2}{K_a}} \tag{5.17}$$

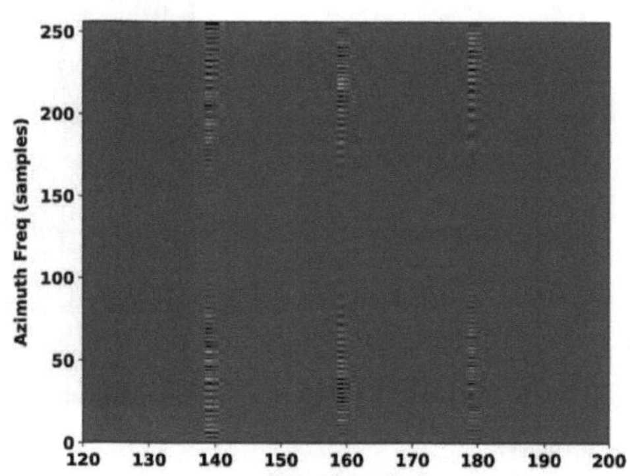

FIGURE 5.10 The difference between $\Re\{S_1\}$ and $\Re\{S_2\}$, zoomed into the scatterer energy area.

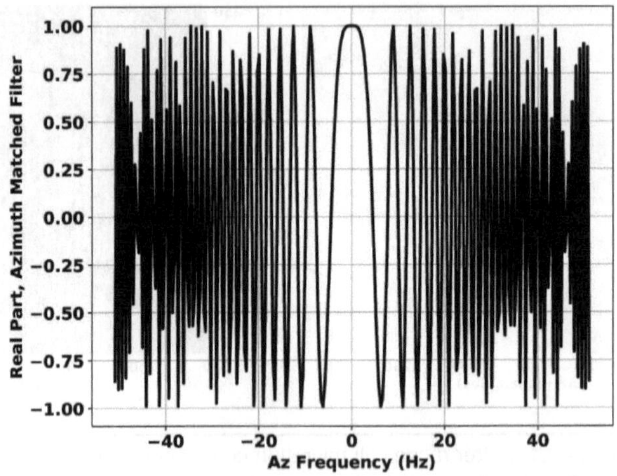

FIGURE 5.11 The azimuth matched filter, applied after RCMC.

Figure 5.11 shows the matched filter for our example. Note that the azimuth bandwidth is determined by the azimuth sampling rate, also known as the *pulse repetition frequency* (PRF). It's clear that the matched filter is the complex conjugate of the second term in Equation (5.16). The matched filter can be applied directly in the range-Doppler domain by multiplying the matched filter across each of the range bins,

$$S_3(t, f_\eta) = S_2(t_j, f_\eta) H_{az}(f_\eta)|_{j=[0, N_r-1]} \tag{5.18}$$

Figure 5.12 shows the results of our three targets after the azimuth matched filter is applied.

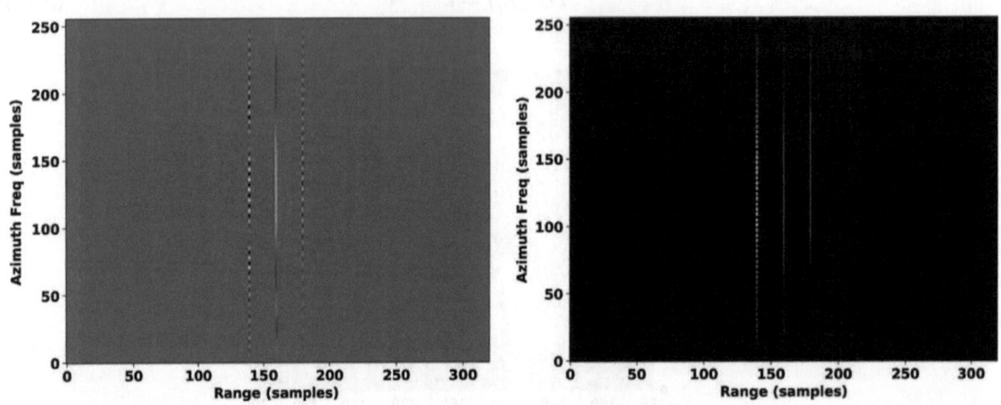

FIGURE 5.12 The data after azimuth matched filter is applied, with the real component of S_3 on the left and the magnitude on the right.

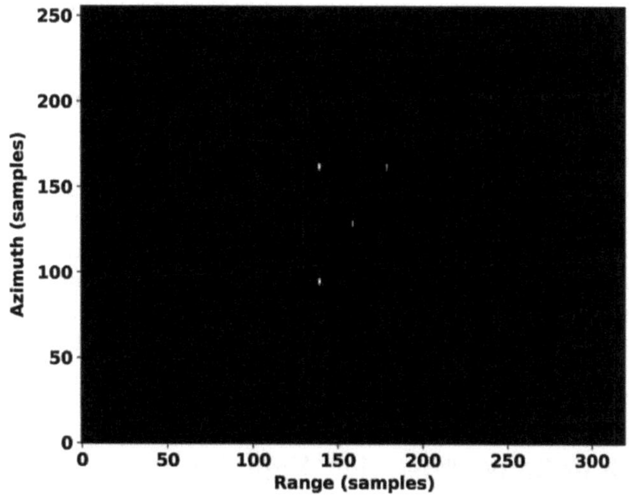

Figure 5.13 The compressed single-look synthetic aperture radar image for the targets shown in Figure 5.2. This magnitude plot has some linear contrast stretch applied to it, making the mainlobes and the sidelobes visible, particularly in the azimuth direction.

5.1.7 Step 5: Azimuth IFFT

After the azimuth matched filter is applied, the range-Doppler data now takes on the form

$$S_3(t, f_\eta) = A_0 p_r \left(t - \frac{2R0}{c}\right) W_a \left(f_\eta - f_{\eta_c}\right) e^{-j\frac{4\pi f_0 R_0}{c}} \tag{5.19}$$

To complete the RDA, an IFFT across each range bin is performed.

$$\begin{aligned} s_{\mathrm{SLC}}(t, \eta) &= \mathrm{IFFT}|_{j=[0,N_r-1]}\{S_3\left(t_j, f_\eta\right)\} \\ &= A_0 p_r \left(t - \frac{2R0}{c}\right) p_a(\eta) \\ &\cdot e^{-j\frac{4\pi f_0 R_0}{c}} e^{j2\pi f_{\eta_c}\eta} \end{aligned} \tag{5.20}$$

This data now resides in the spatial domain.[7] The p_r term remains and a second envelope p_a has been added. These two envelopes represent a two-dimensional sinc-like function varying in range and azimuth for each target. As written, these terms shown that the target is centered at $t = 2R_0/c$ and $\eta = 0$, where R_0 is the closest approach for a given scatterer and η is relative to the time of closest approach.

Figure 5.13 shows the focused, compressed single-look SAR magnitude image after RDA. This image is produced by taking the magnitude of S_3 using something like `abs` in either MATLAB or Python Numpy. Note that the plot has some linear contrast stretch associated with it, giving the targets a slight width in azimuth. Further image quality

[7] This terminology is often confusing to some people, as the dimensions are in time, but it's called spatial. One is a dual of the other in radar, since time is literally a measure of range, which is a spatial quality.

processing, such as aperture weighting and upsampling, would correct for this and provide a crisp, sharp magnitude image.

It is worth noting that this description of the RDA is only valid for small aperture sizes and low-squint angles. There are other variations of this algorithm (see, e.g., Jin and Wu, 1984[8]) that correct for large range migration due to either wide apertures or high-squint cases.

The implementation of the RCMC [and of the secondary range correction in the case of high-squint and/or wide apertures] makes the implementation of the RDA under some conditions either difficult to implement or may provide high computing loads. There are other algorithms that assist in correcting the range migration in a more efficient fashion. One of these, called the Omega-K (ω-K) or RMA, is able to accurately process data over wide apertures and/or high-squint angles. It will be discussed in the next section.

5.2 Omega-K or Range Migration Algorithm

The RMA was first proposed by Cafforio, Prati, and Rocca in 1991.[9] The technique derives from a seismic signal processing technique described by Stolt.[10]

Seismic imaging is performed using a set of geophones and a detonation charge. The geophones act as the aperture, while the detonation charge becomes the illumination source. Once the charge is set off, the geophones record the echoes from the geological features under the ground. Since pulsed SAR treats each pulse as if the moving platform was "stopping and hopping," the geophones become analogous with this stop-n-hop philosophy. That is, there is a geophone in each location across the aperture to record the incoming seismic wave, just as the SAR listens for a set of returns for each transmitted pulse across the synthetic aperture.

Stolt developed a frequency domain solution to locate the features in the returned seismic data. This highly accurate technique is called Stolt mapping and has been used in SAR image formation, as well as in medical imaging.[11]

Cumming and Wong[12] have a comprehensive description of the Omega-K in Chapter 8. Carrara, Goodman, and Majewski[13] have a chapter outlining Omega-K, which they refer to as range migration, for dechirped data in a spotlight geometry. A summary of the stripmap, demodulated version of the algorithm will be given here.

The first step in the algorithm, which is shown in Figure 5.14, is a two-dimensional (2D) FFT that puts the data into both the range and azimuth frequency domain. From here, a *reference function multiply* or *bulk compression* is applied to the frequency domain

---

[8] M. J. Jin and C. Wu, "A SAR Correlation Algorithm Which Accommodates Large Range Migration," *IEEE Trans. Geoscience and Remote Sensing*, Vol. 22, No. 6, November 1984.

[9] C. Cafforio, C. Prati, and F. Rocca, "SAR Data Focusing Using Seismic Migration Techniques," *IEEE Trans. on Aerospace and Electronic Systems*, Vol. 27, No. 2, March 1991.

[10] R. H. Stolt, "Migration by Transform," *Geophysics*, Vol. 43, No. 1, February 1978.

[11] D. Garcia, L. L. Tamec, S. Muth, E. Montagnon, J. Poree, and G. Cloutier, "Stolt's f-k Migration for Plane Wave Ultrasound Imaging," *IEEE Trans. Ultrasonics, Ferroelectrics, and Frequency Control*, Vol. 50, No. 9, September 2013.

[12] I. G. Cumming and F. H. Wong, *Digital Processing of Synthetic Aperture Radar: Algorithms and Implementation*, Artech House, Boston, MA, 2005.

[13] W. G. Carrara, R. S. Goodman, and R. M. Majewski, *Spotlight Synthetic Aperture Radar: Signal Processing Algorithms*, Artech House, Boston, MA, 1995.

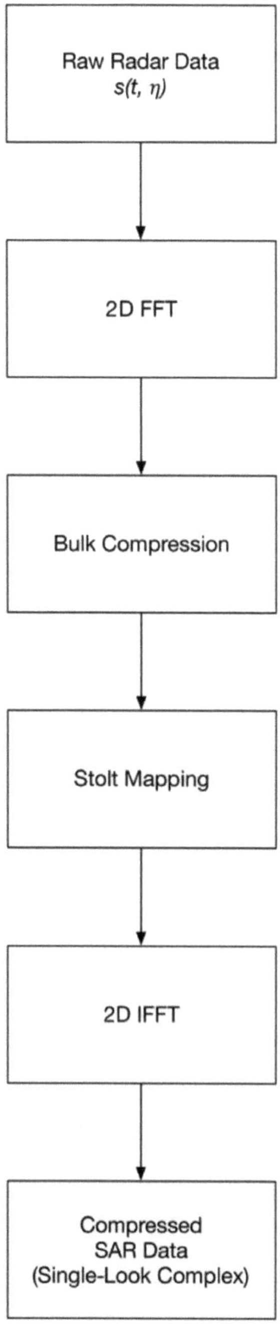

FIGURE 5.14 The algorithm flow for Omega-K/Range Migration Algorithm.

data that compresses the data in both range and azimuth simultaneously. The bulk compression is typically referred in range at the scene center. Targets at that range will be focused, but the other targets in the scene will only be partially focused. The reference function is defined to be

$$\theta_{\text{ref}}(f_t, f_\eta) = \frac{4\pi R_{\text{ref}}}{c}\sqrt{(f_0 + f_t)^2 - \frac{c^2 f_\eta^2}{4v_r^2}} + \frac{\pi f_t^2}{K} \tag{5.21}$$

where R_{ref} is the selected reference range (typically at the GRP). The reference range is defined in the range time domain, rather than in the frequency domain, as the range variation cannot be easily described in the frequency domain.

The 2D range/azimuth frequency domain data is represented by

$$\begin{aligned} S_{\text{qd}}(f_t, f_\eta) &= AW_r(f_t)W_a(f_\eta, -f_{\eta_c}) \\ &\cdot e^{-j\frac{4\pi R_0}{c}\sqrt{(f_0+f_t)^2 - \frac{c^2 f_\eta^2}{4v_r^2}} + \frac{\pi f_t^2}{K}} \end{aligned} \tag{5.22}$$

After the bulk compression, the radar signal takes on the form

$$\begin{aligned} S_{\text{RFM}}(f_t, f_\eta) &= AW_r(f_t)W_a(f_\eta, -f_{\eta_c}) \\ &\cdot e^{-j\frac{4\pi(R_0 - R_{ref})}{c}\sqrt{(f_0+f_t)^2 - \frac{c^2 f_\eta^2}{4v_r^2}}} \end{aligned} \tag{5.23}$$

Cumming and Wong remind us that this approximation is valid because v_r is independent of range.

The next step in RMA is to perform the *Stolt mapping*, which fully focuses the targets not at the reference range. The mapping interpolates each range bin by mapping f_t to f_t',

$$f_t' = \sqrt{(f_0 + f_t)^2 - \frac{c^2 f_\eta^2}{4v_r^2}} - f_0 \tag{5.24}$$

This mapping essentially shifts and scales the targets in frequency space. Each range frequency bin is interpolated onto the new frequency mapping. Figure 5.15 illustrates the effect of the mapping on the radar data. The data is warped into a shape reminiscent of the change in range during the dwell (i.e., it's parabolic in shape). The data in the center of the dwell, which corresponds to R_{ref}, are essentially left untouched in the mapping. This is desired, as they are already focused.

After the mapping, the data in the range/azimuth frequency domain becomes

$$\begin{aligned} S_{\text{Stolt}}(f_t', f_\eta) &= AW_r(f_t)W_a(f_\eta, -f_{\eta_c}) \\ &\cdot e^{-j\frac{4\pi(R_0 - R_{ref})}{c}(f_0 + f_t')} \end{aligned} \tag{5.25}$$

which is linear in the new range frequency domain. The higher order terms have been removed. A 2D inverse FFT is performed and a fully compressed, fully focused single-look complex image is the result.

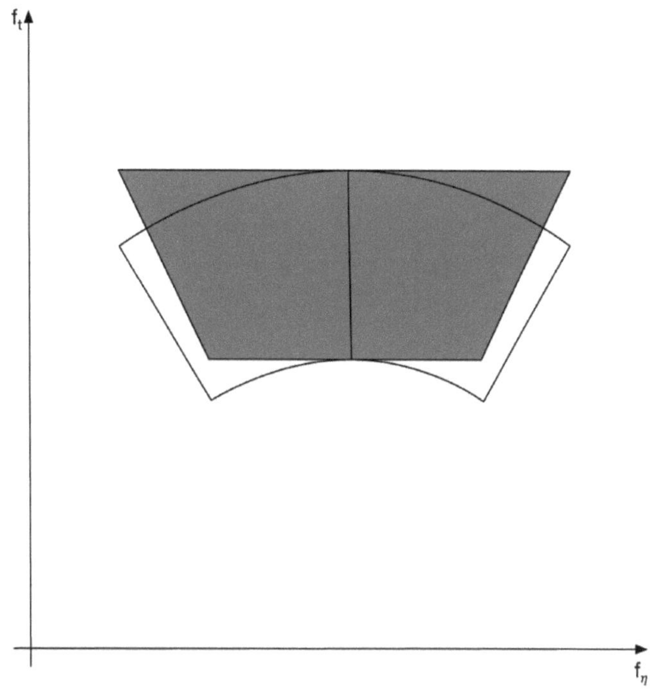

FIGURE 5.15 The effect of Stolt mapping on the radar data.

5.3 Quadrature Demodulation versus Dechirp

As has been previously mentioned, the designation that these algorithms are for quadrature demodulated data is somewhat artificial. These algorithms are well-suited for focusing data that has been demodulated with a single sinusoid, rather than a reference chirp, but with minor modifications, some of these algorithms can be used with dechirped data as well. Conversely, some of the algorithms that are described in the next chapter can be used with quadrature demodulated data as well. In selecting an algorithm, processing load, memory requirements, phase accuracy after compression, imaging geometry, and processing times are the trade-offs that need to be considered. For example, does an algorithm require storing blocks of phase history to process or can the process be performed pulse-by-pulse and/or line-by-line? Are real-time results required? If so, does an approximately focused image meet requirements? These are just some of the questions that need to be asked when considering which image formation algorithm to implement.

CHAPTER 6

Image Formation Algorithms for Dechirped Data

Here a comparison of quadrature demodulation and dechirp techniques will be reviewed. The advantages and disadvantages of each of the techniques will be discussed. The mathematical form of the dechirped signal will be revisited, pulse by pulse. Two image formation algorithms focused on dechirped data will be presented: Polar Formatting Algorithm (PFA) and Backprojection (BP).

6.1 Polar Formatting

As you will recall from Chapter 3, the advantage of performing dechirp (sometimes called stretch processing) on the received radar data is that the matched filtering is done during the dechirp process, leaving behind a sinusoid at a unique frequency for each quantized range bin in the scene. This is shown graphically in Figure 6.1. There are two plots in this figure; the top plot represents the radar signals from three different scatterers, one near the edge of the swath (N), one at the scene center (C), and one at the far edge of the swath (F). The plot is in time-frequency space, where the independent axis is fast time and the dependent axis is frequency. A chirp is transmitted with some width T_p shown as the horizontal line at the far left of the top plot. After some round-trip time delay dependent on the range of the target to the sensor, each of the chirps from N, C, and F comes back to the sensor. Target N arrives first, with Target C arriving second some amount of time after N, and then Target F arriving last. The signals are mixed with a reference chirp and what is left are three different sinusoids, represented by the flat lines in the bottom plot. Each has a unique frequency. The width between the near edge returns and the far edge returns are called the video bandwidth B_v. It is worth noting that B_v is much less than B, significantly reducing the sampling rate for analog-to-digital conversion. The time delay that each target takes to get to the sensor is still present after dechirp, which is why each of the flat lines is staggered. This is called *range skew*.

If you couple dechirp with motion compensation, the mixing operation removes the chirp in both range and azimuth, transforming the range-time domain to the range-frequency domain (again, a unique frequency sinusoid for each range bin) and the

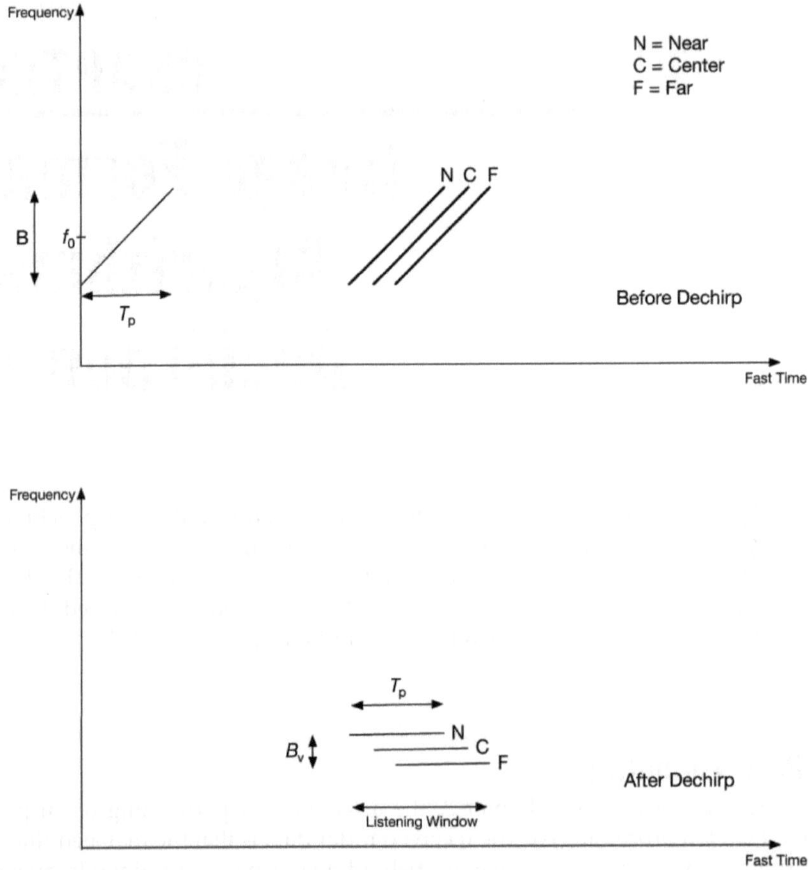

FIGURE 6.1 Graphical illustration of dechirping.

azimuth-time domain to the azimuth-frequency domain (a unique frequency sinusoid for each azimuth bin).[1] Thus, a single 2D FFT places the dechirped/motion compensated data into the range-azimuth spatial domain.

This seems ideal, making for a much easier, much faster way to focus data. However, there is still one issue that has to be resolved: the range cell migration (RCM) problem. The data collected still has the range migration in it spatially. The data is collected on an annulus. Figure 6.2 shows how the data is arranged in the wavenumber (or range-azimuth frequency) space. The data ends up being on an annulus, where the bounds are ΔK along K_x (range-frequency) and $\Delta\theta$ along K_y (azimuth-frequency). The digitized data is saved virtually in a rectangular grid and, for a 2D Fourier Transform, is referenced on that same equally spaced rectangular grid, the 2D sinusoids that represent each scatterer in the scene are deformed the further away you get from K_c and θ_{ac}, which is the center

[1] W. G. Carrara, R. S. Goodman, R. M. Majewski, *Spotlight Synthetic Aperture Radar: Signal Processing Algorithms*, Artech House, Boston, MA, pp. 67, 99, 1995.

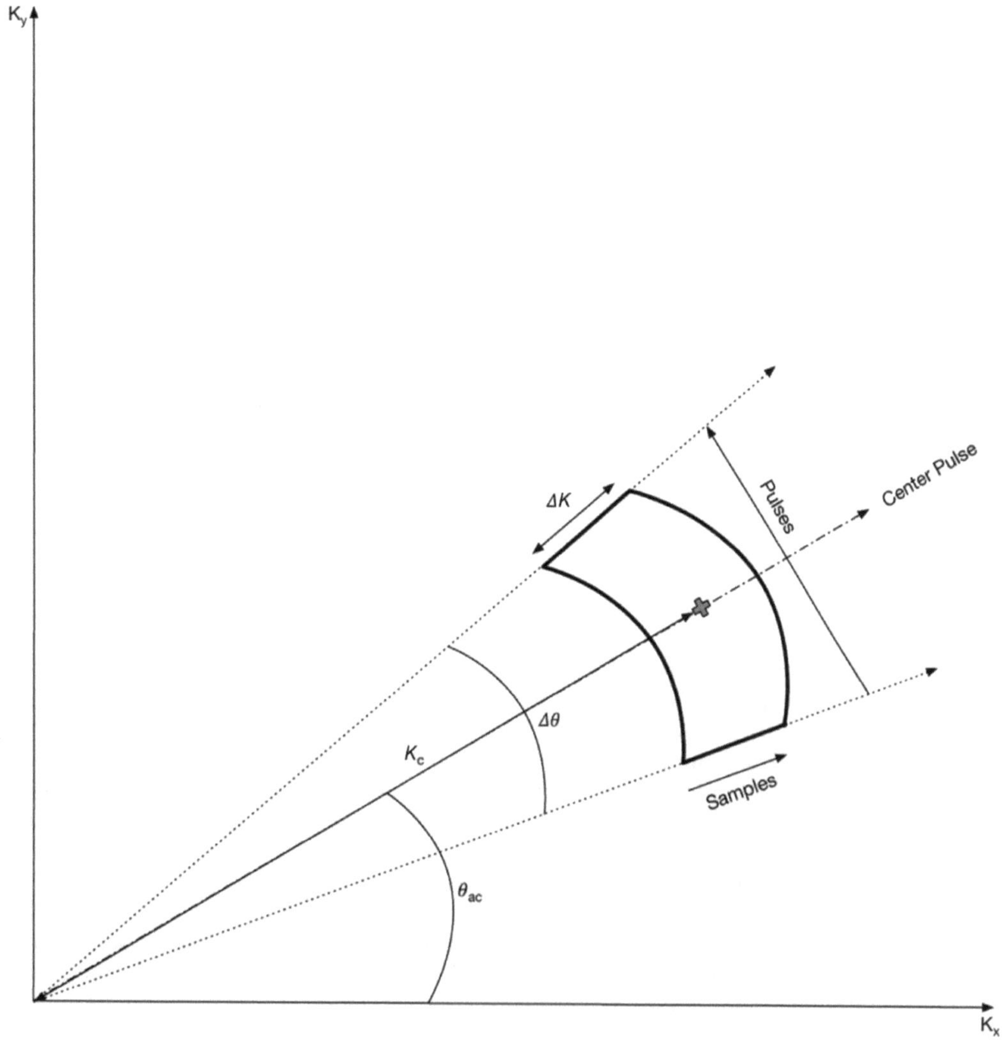

FIGURE 6.2 The data collection geometry in wavenumber (or frequency) space.

point of the data collection annulus (shown in gray in the figure). Thus, points that are further from the scene center are more and more out of focus.

Polar formatting deals with this by placing a rectangular grid over top the polar grid, either circumscribing or inscribing the polar grid, and interpolating across the entire grid. This then puts the recorded samples in the proper coordinate system, removes the distortion, and focuses the scatterers across the entire aperture. If the data is circumscribed, the rectangular grid is larger than the polar format grid. Those areas that fall outside the polar grid are typically set to zero. If the data is inscribed, the rectangular grid falls within the boundary of the polar grid and some data on the edges is discarded or ignored. Circumscribed data maintains its full azimuth resolution, while some minor loss of azimuth resolution may occur for inscribed grids, particularly for wide apertures.

The center of the dwell is represented by θ_{ac}. The center of the samples is represented by K_c and can be calculated by

$$K_c = \frac{4\pi f_0}{c} \tag{6.1}$$

where f_0 is the center frequency of the radar and c is the speed of light. The total width of the samples ΔK is calculated as

$$\Delta K = \frac{4\pi B}{c} \tag{6.2}$$

where B is the transmitted bandwidth.

Figure 6.3 describes the algorithmic steps necessary to perform the PFA. Assuming that the data has been dechirped and motion compensated, the radar signal now has the form of

$$\begin{aligned} s_{dc}(\eta, t) = Bw_a(\eta)\, \mathrm{rect}\left(\frac{t - (\tau(\eta) - \tau_c(\eta))}{T}\right) \\ \cdot e^{j2\pi K(\tau(\eta)-\tau_c(\eta))t + 2\pi f_0(\tau(\eta)-\tau_c(\eta)) - \pi K(\tau(\eta)^2 - \tau_c(\eta)^2)} \end{aligned} \tag{6.3}$$

where τ represents the round-trip time delay to some arbitrary scatterer in the scene and τ_c is the round-trip time delay to the scene center. Note that the term $-\pi K(\tau(\eta)^2 - \tau_c^2)$ is called the *residual video phase* (RVP). When the aperture is sufficiently large, this term cannot be ignored and must be compensated. This term is deterministic and can be easily removed before polar formatting is applied. Carrara et al.[2] discuss the removal of RVP as part of range deskew in their Appendix C. For simplicity, we will ignore this term, simulating it but not correcting for it. We can safely ignore it as part of the example because our swath size is small enough to make the term negligible.

As an example to help illustrate how PFA works, consider the radar parameters in Table 6.1. This table is very similar to the parameters shown in Chapter 5, with some notable exceptions. First, the initial and final squint angles are identified, indicating that this is a spotlight mode collect, rather than a stripmap mode as was the previous example. The change in squint is performed either mechanically (e.g., steering the antenna with a gimbal) or electronically (e.g., beamforming). Second, the grazing angle is identified, indicating the angle between the slant plane and the ground plane at the center of the dwell. With this, we can use trigonometry to calculate the slant range from antenna to the ground plane at the center of the dwell. The number of azimuth lines and range samples is also different, due to the change in imaging mode. Finally, the range and azimuth resolution are indicated. These are given, respectively, by $\rho_r = c/(2B)$ and $\rho_a = \lambda/(2\Delta\theta)$, where $\Delta\theta$ is to total squint over the dwell. Finally, the slant range to the GRP is measured at the center of the dwell.

Figure 6.4 illustrates a simulated scene where the four targets are placed around and at the GRP. The range from the antenna to the GRP is shown in Figure 6.5. You will notice

[2] W. G. Carrara, R. S. Goodman, and R. M. Majewski, *Spotlight Synthetic Aperture: Signal Processing Algorithms*, Artech House, Boston, MA, pp. 501–506, 1995.

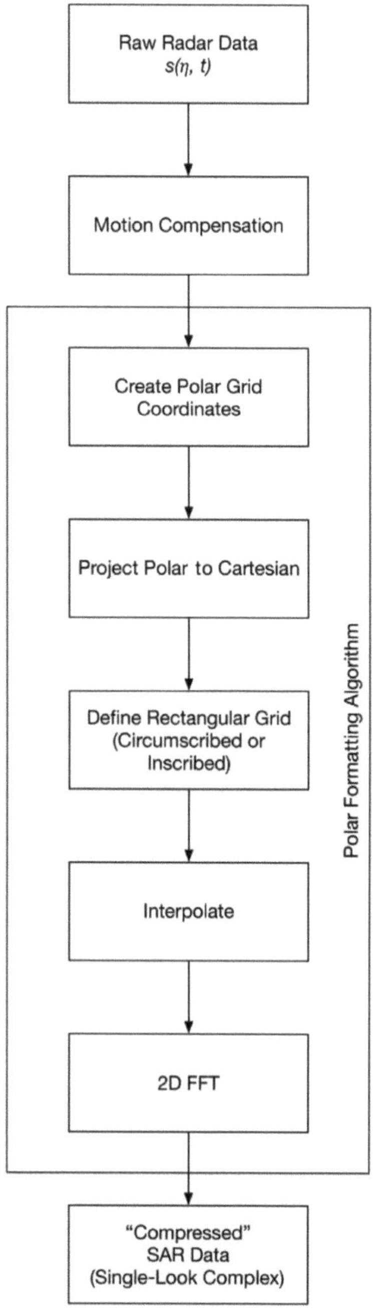

FIGURE 6.3 The algorithm flow for the Polar Formatting Algorithm (PFA)

that the target placement is the same as the example in Chapter 5, while the range to GRP is nearly the same, the biggest difference being the total dwell time, which is more than double the previous example.

Using Equation (6.3), raw radar data after motion compensation is generated. The magnitude and phase of this data is shown in Figure 6.6.

Parameter Name	Variable	Value/Units
Radar platform height	z_a	20 km
Radar velocity	v_a	150 m/s
Grazing angle	ψ_b	45°
Initial squint angle	θ_i	−1.0°
Final squint angle	θ_f	+1.0°
Pulse duration	T	6.8 μs
Range chirp rate	K	20 MHz/μs
Center frequency	f_0	5.3 GHz
Range sampling rate	F_r	60 MHz
Azimuth sampling rate	F_a	100 Hz
Number of range lines	N_a	658 lines
Number of samples per range line	N_r	449 samples
Range resolution	ρ_r	1.0 m
Azimuth resolution	ρ_a	0.81 m

TABLE 6.1 Parameters for C-Band Airborne Synthetic Aperture Radar

If s_{dc} is transformed into frequency space using a 2D FFT, the resulting image will contain all of the scatterers. Figure 6.7 displays the four scatterers for our case. As the scatterers get further and further away from the center of the scene, the more distorted they become. This distortion is due to range cell migration. Polar formatting will correct for this; however, as the aperture gets wider, polar formatting will not compensate for

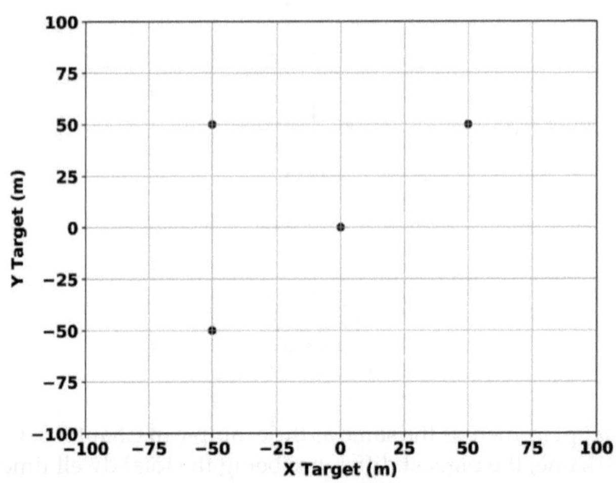

FIGURE 6.4 Four scatterers placed in a simulated scene. The antenna beam pattern is steered during the dwell so that the GRP is always in the center of the beam.

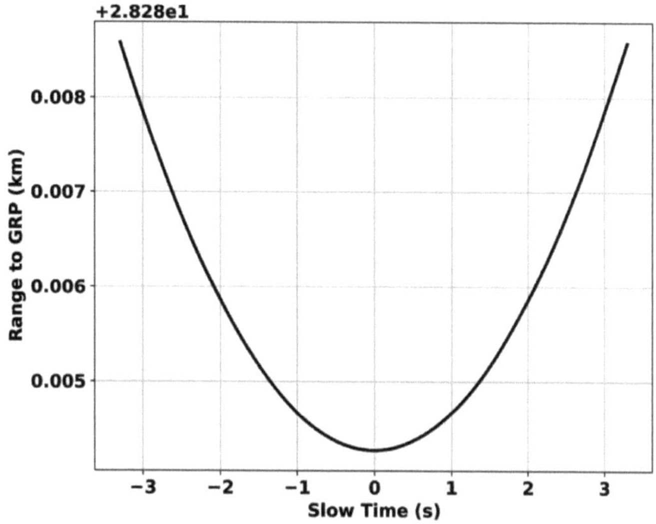

Figure 6.5 Range to the GRP from the antenna over all of slow time.

it perfectly and a technique like ω-K or the BPA, which we'll discuss in the next section, will have to be used.

With $\Delta\theta$, the total squint over the dwell, and ΔK, K_c, the samples in K-space, a polar grid is effectively defined, and the polar grid can be converted to Cartesian coordinates. Looking carefully at Figure 6.6, it's clear that the data are distorted, curved around the azimuth or slow time axis like an annulus.

6.1.1 Steps 1 and 2: Create Polar Grid and Project Polar-to-Cartesian

The first step in polar formatting is to define that annulus space. Using evenly spaced values for $\Delta\theta$, we can then define the indices in K_x. Using Equations (6.1) and (6.2), we can define the indices in K_y space. Two matrices can be created for $\Delta\theta$ and ΔK (as in Figure 6.2). These coordinates represent the data as it exists in polar space. Figure 6.8 shows the radar data coordinates in a polar plot.

In reality, the data is stored in a digital rectangular grid and see by the Fast Fourier Transform as a rectangular grid. So the next step is to map the data from a polar coordinate system to Cartesian or rectangular coordinate system. This is done by

$$\begin{aligned} K_x &= r\cos\left(\theta\right) \\ K_y &= r\sin\left(\theta\right) \end{aligned} \tag{6.4}$$

where $r = [K_c - (\Delta K/2), K_c + (\Delta K/2)]$ and $\theta = [-\Delta\theta/2, \Delta\theta/2]$. Figure 6.9 shows the coordinates after mapping from polar to Cartesian. Note the slight range curvature that exists along the K_y axis. This is the space our data currently lives in: not quite rectangular. As the squint angle moves off broadside or the closer the ranges get to sensor, the more distorted this rectangle becomes. In essence, the distortion follows the range isolines as they fall in the ground plane. The worse the distortion, the greater the need

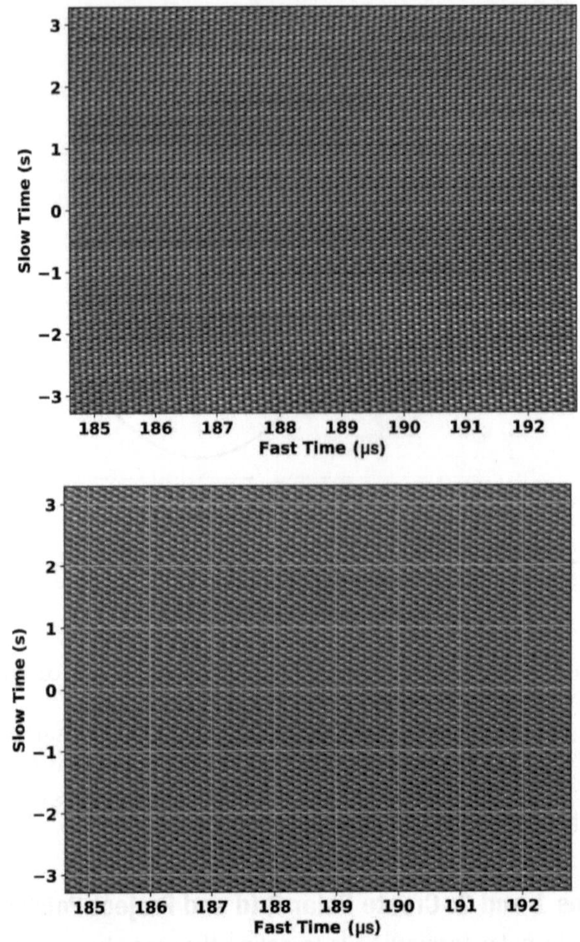

FIGURE 6.6 The magnitude (top) and phase (bottom) of the simulated dechirp radar data for the four scatterers.

for compensation. That is, in some cases, PFA might not be needed if the scene is very small or the ranges are far enough away so that the range isolines are essentially linear, provided that the aperture is sufficiently small.

6.1.2 Step 3: Define Inscribed or Circumscribed Grid

The next step requires a grid be defined. This rectangular grid either inscribes or circumscribes the data, as illustrated in Figure 6.10. The data is then interpolated onto that grid using a 2D interpolation scheme. Both Jakowatz et al. (1996)[3] and Carrara et al. (1995)[4]

---

[3] C. V. Jakowatz, D. E. Wahl, P. H. Eichel, D. C. Ghiglia, and P. A. Thompson, *Spotlight-Mode Synthetic Aperture Radar: A Signal Processing Approach*, Kluwer Academic Publishers, Boston, MA, pp. 133–156, 1996.

[4] W. G. Carrara, R. S. Goodman, and R. M. Majewski, *Spotlight Synthetic Aperture Radar: Signal Processing Algorithms*, Artech House, Boston, MA, pp. 180–188, 1995.

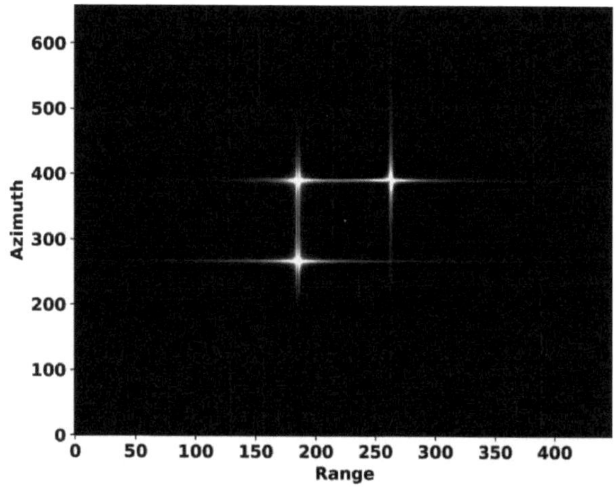

Figure 6.7 Four scatterers in the scene before the Polar Formatting Algorithm (PFA) is applied. Note that the further away from the scene center, the more distorted the targets become; the target at the GRP is perfectly focused, while the other three targets are slightly smeared in both azimuth and range.

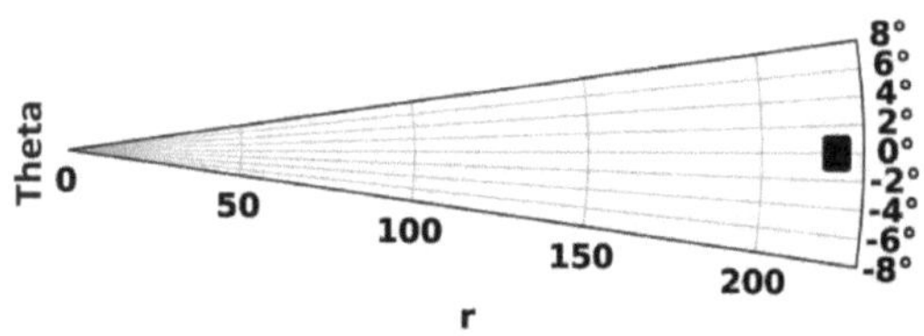

Figure 6.8 Coordinates of s_{dc} shown in polar coordinate space. Note that the image, shown in black, is square in this space.

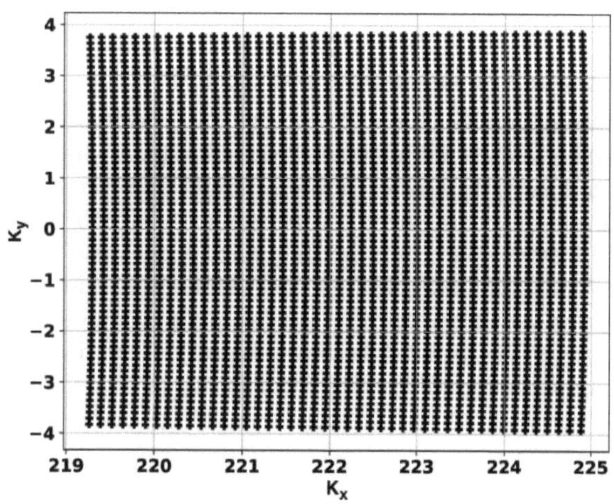

Figure 6.9 The Cartesian coordinate indices for the s_{dc}.

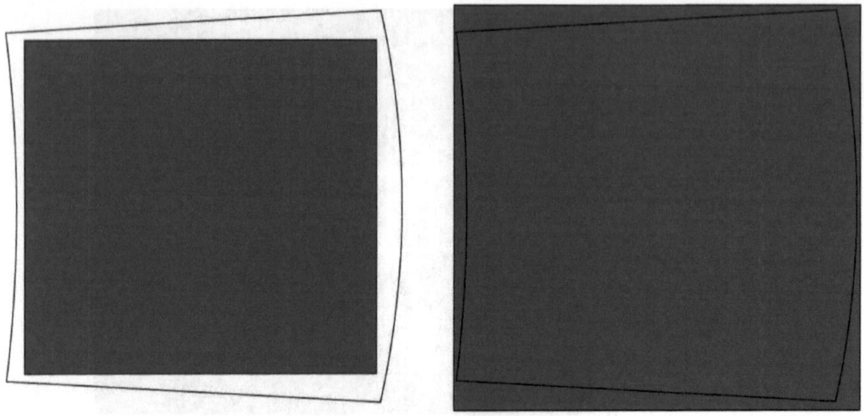

Figure 6.10 Illustration showing an inscribed rectangle (left) or a circumscribed rectangle (right).

describe interpolation techniques that interpolate the data along each pulse and then along each range bin. Since memory and computing power is readily available nowadays, this scheme is not necessary, as there are sufficient 2D interpolation techniques that can work on the entire aperture at once. If the aperture is very wide, the techniques described by either of these sources may be needed to manage memory and computational complexity. With the advent of Graphical Processing Units (GPUs), fast interpolation should not be difficult nor tedious.[5] Care should be taken that whatever 2D algorithm is used to interpolate the data, it must be able to handle nonuniform grid data as its input (as our data lives on an annulus, which is evenly spaced in polar coordinates, but not in Cartesian). For this example, an inscribed rectangle is selected. This is shown in Figure 6.11.

6.1.3 Steps 4 and 5: Interpolate and 2D FFT

The data is interpolated using a 2D interpolation technique. The newly interpolated magnitude data is shown in Figure 6.12. Note that the distortion has been removed, the data appears to be evenly spaced in the rectangle, and that data spans an annulus. The range cell migration has now been compensated. A 2D FFT will then take the data from the raw phase history space into the frequency space, also known as the image space. Figure 6.13 compares the magnitude image data before and after the application of the PFA. The comparison illustrates the improvement in image quality after PFA (tightening of each scatterer's IPR). Sidelobes are still present, but the mainlobe has been shrunk, providing a sharper image. The reduction of scatterer IPR sidelobes will be discussed in Chapter 8.

[5] For example, see P. K Agarwal, A. Beutel, T. Molhave, "TerraNNI: Natural Neighbor Interpolation on 2D and 3D Grids Using a GPU," *ACM Transactions on Spatial Algorithms and Systems*, Vol. 2, No. 2, Article 7, June 2016.

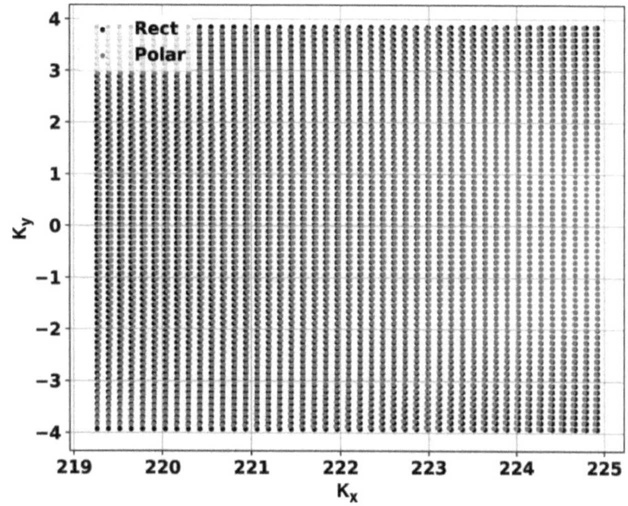

Figure 6.11 Inscribed rectangle coordinates for interpolation. The original Cartesian (from polar) coordinates are shown in gray and the new evenly spaced, inscribed rectangular coordinates are shown in black. Note that the distortion along each range bin does not appear to be significant, but is enough to cause the distortion shown in Figure 6.7.

6.2 The Backprojection Algorithm

Computer-aided tomography (CAT) is a well-known technique in the medical field that uses X-ray imaging over varying viewing angles to create an unobscured view of various bodily organs.[6] This reduces or eliminates shadowing from other organs within the body. The geometric model for collecting CAT scans is similar to synthetic aperture radar (SAR).[7] One might be tempted to assume that this technique is only applicable to spotlight-mode geometry, but this is not the case; with minor modifications, tomographic processing can be used with stripmap mode SAR as well. The biggest difference between SAR and CAT is that the returned energy from the imaging scene comes along a parallel path from the X-ray emitter to the detector, where as the path from the transmitter to the receiver is a perpendicular path (due to the reflection of the RF energy, rather than the attenuation of the X-ray radiation as in CAT imaging).

The processing engine for this technique is called the Backprojection Algorithm, sometimes referred to as "convolutional backprojection," or, simply, "tomography."[8,9]

[6] H. J. Scudder, "Introduction to Computer Aided Tomography," *Proceedings of the IEEE*, Vol. 66, No. 6, June 1978.

[7] D. C. Munson, J. D. O'Brien, and W. K. Jenkins, "A Tomographic Formulation of Spotlight-Mode Synthetic Aperture Radar," *Proceedings of the IEEE*, Vol. 72, No. 8, August 1983.

[8] J. L. Bauck and W. K. Jenkins, "Tomographic Processing of Spotlight-Mode Synthetic Aperture Radar Signals with Compensation for Wavefront Curvature," *IEEE 1998 International Conference on Acoustics, Speech, and Signal Processing*, New York, NY, 11–14 April 1988.

[9] M. D. Desai and W. K. Jenkins, "Convolutional Backprojection Image Reconstruction for Spotlight Mode Synthetic Aperture Radar," *IEEE Transactions on Image Processing*, Vol. 1, No. 4, October 1992.

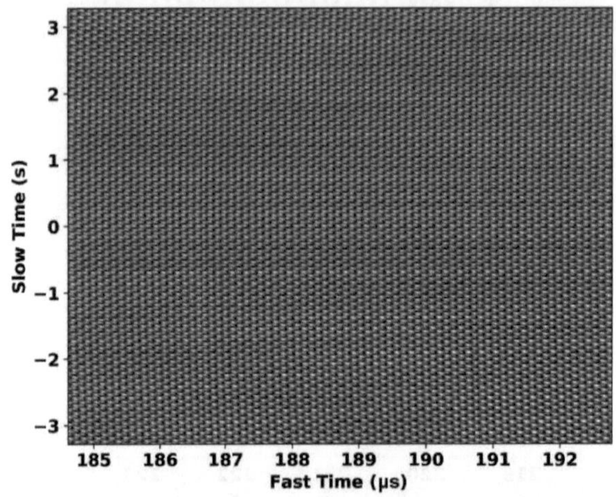

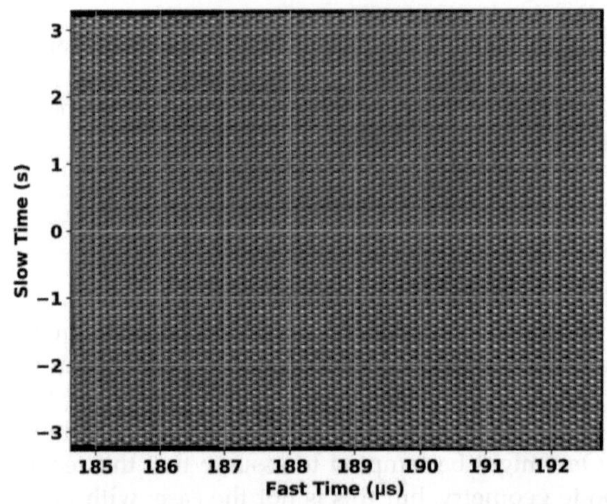

FIGURE 6.12 Comparison of the magnitude of s_{dc} before and after the PFA. The magnitude data before PFA is shown at top, while the magnitude data after PFA is shown at bottom. Note the distortion has been removed and the data appears to be evenly spaced in the shape of an annulus.

Backprojection relies on the *projection-slice theorem*.[10] The general idea behind this theorem is that the Fourier transform of the projection at some angle θ is a slice of the 2D Fourier transform of the entire scene, taken at that angle θ. That is, the entire SAR (or CAT) scene is represented in each pulse (or slice, if you will). Given enough of these slices, the scene can be reconstructed. Given an infinite number of these slices, spaced over an entire circle, the scene could be perfectly reconstructed. This hints to the holographic nature of this type of imaging, which we will cover in detail in Chapter 8.

[10] G. T. Herman, *Image Reconstruction from Projections*, Springer, New York, NY, 1979.

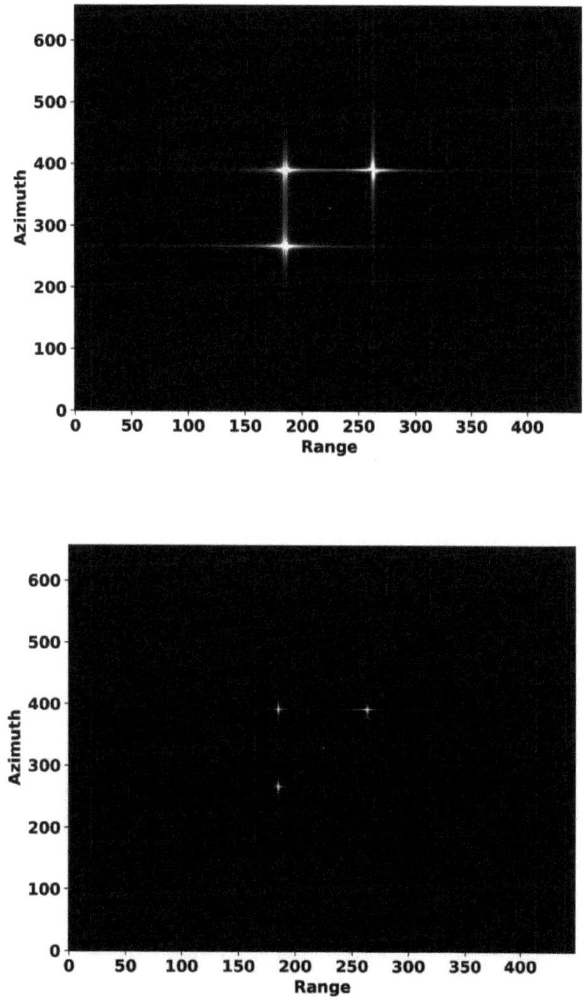

FIGURE 6.13 A comparison of the image data before PFA (top) and after PFA (bottom). Note that all the targets are focused after PFA, the target at the GRP was essentially untouched, while the targets away from the GRP are focused.

6.2.1 Computer-Aided Tomography

In CAT, X-ray emitters are positioned in parallel with detectors, leaving space in between the emitters and detectors for the object to imaged. The emitters release pulses of radiation and the detectors measure the incoming radiation from the emitters. The object attenuates the radiation by some factor $g(x, y)$. The detectors measure the intensity of the incoming energy, the projections of g which are found using

$$p_\theta(u) = -\log \frac{I_\theta(u)}{I_0} \tag{6.5}$$

where I_0 is the intensity of the original X-ray source and $I_\theta(u)$ is the received intesity at the detector. Note that the received X-ray beam follows an exponential dependence on the line integral of g. $p_\theta(u)$ is a projection that is obtained by changing θ by rotating either the object or, more likely, the array of sources and detectors.

There are several methods by which the CAT image can be reconstructed, including the convolutional backprojection method.[11] The method described here will be the direct Fourier domain reconstruction. It is suitable for both SAR and medical tomography.[12]

Direct Fourier domain reconstruction is dependent on the Projection Slice Theorem.[13] Let $g(x, y)$ be an unknown signal (either attenuation in CAT or radar reflectivity in SAR). The Fourier transform of g is

$$G(X, Y) = \int_{-\infty}^{+\infty} \int_{-\infty}^{+\infty} g(x, y) e^{-j(xX+yY)} dxdy \tag{6.6}$$

The inverse Fourier transform is thus defined as

$$g(x, y) = \frac{1}{4\pi^2} \int_{-\infty}^{+\infty} \int_{-\infty}^{+\infty} G(X, Y) e^{+j(xX+yY)} dXdY \tag{6.7}$$

If $p_\theta(u)$ is defined to be a projection of g at an angle θ, as shown in Figure 6.14, then that projection is mathematically defined as

$$p_\theta(u) = \int_{-\infty}^{+\infty} g(u\cos(\theta) - v\sin(\theta), u\sin(\theta) + v\cos(\theta))dv \tag{6.8}$$

where p_θ is parallel to the v axis at some location u, and each $u = u_n$, where n is a given emitter/detector pair, which are parallel to the u axis.

The 1D Fourier transform of $p_\theta(u)$ is defined to be

$$P_\theta(U) = \int_{-\infty}^{+\infty} p_\theta(u) e^{-juU} du \tag{6.9}$$

where U is the slice projection at $(U\cos(\theta), U\sin(\theta))$. The projection slice theorem is more commonly stated as

$$P_\theta(U) = G(U\cos(\theta), U\sin(\theta)) \tag{6.10}$$

This equates to the a slice of the 2D transform $G(X, Y)$ taken at an angle θ with respect to the X-axis. We will use this relationship to reconstruct the scene (or object, in the case of CAT) by "backprojecting" the data onto the 2D $G(X, Y)$ grid.

---

[11] H. J. Scudder, "Introduction to Computer Aided Tomography," *Proceedings of the IEEE*, Vol. 66, No. 6, June 1978.

[12] H. H. Stark, J. W. Woods, I. Paul, and R. Hingorani, "Direct Fourier Reconstruction in Computer Tomography," *IEEE Trans. Acoustics, Speech, and Signal Processing*, Vol. ASSP-29, April 1981.

[13] R. N. Bracewell, "Numerical Transforms," *Science*, Vol. 248, No. 4956, 11 May 1990.

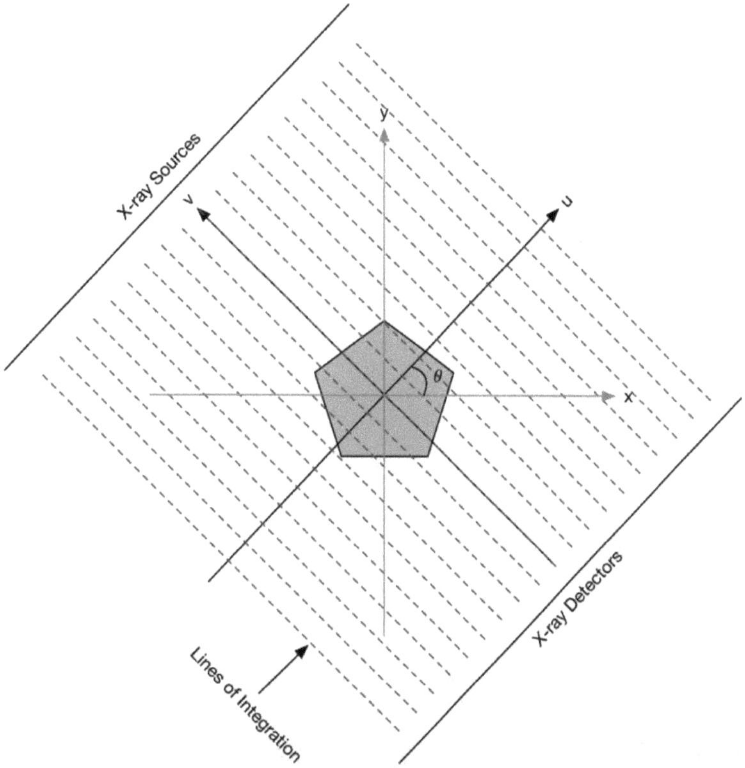

Figure 6.14 A cartoon illustration of computer-aided tomography (CAT). The object, shown as a polygon in light gray in the center of the illustration, has an unknown X-ray intensity of $g(x, y)$.

6.2.2 Synthetic Aperture Radar Tomographic Model

The SAR tomographic model is similar to the CAT model, except that the projected slice $p_\theta(u)$ is perpendicular to emitted energy, rather than parallel with it. Again, while this is easier to visualize for spotlight mode, the concept is applicable to stripmap as well.

There are a series of integration lines (equivalent to our range lines from previous chapters). The energy for each scatterer in the illuminated patch of ground (shown as a gray circle in Figure 6.15) is integrated along that line and the collection of those samples comprises $p_\theta(u)$. The width of the ground patch is $2L$ and the squint angle is defined by θ. The new axes are parallel and orthogonal to the axis created by the squint angle θ, which we will label as (u, v).

As was described previously, CAT uses the intensity of the measured radition as the projection directly. In SAR, since we are using a chirp waveform and dechirping the data, the measured energy is the transform of the projection. It is not difficult to switch back and forth between the projection and its transform. Additionally, CAT typically images the entire 360° field. SAR typically does not, making SAR a narrowband version of this tomographic method. Despite its narrowband nature, it does not take a good deal of θ angles to create a high-resolution, high-quality image. Recall that azimuth resolution is defined by $\rho_a = \lambda/2\Delta\theta$.

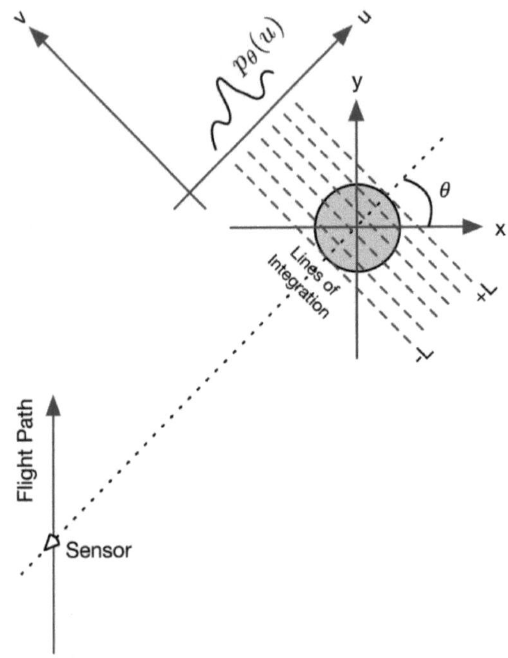

FIGURE 6.15 A cartoon illustration of the tomographic model for SAR.

6.2.3 Reconstruction Using the Backprojection Algorithm

The BPA starts with taking Equation (6.7) and putting it into polar coordinates. Recalling that $x = r\cos(\theta)$ and $y = r\sin(\theta)$,

$$g(\rho\cos(\Theta), \rho\sin(\Theta)) = \frac{1}{4\pi^2}\int_{-\pi/2}^{+\pi/2}\int_{-\infty}^{+\infty} G(r\cos(\theta), r\sin(\theta)) \cdot |r|e^{+jr\rho\cos(\Theta-\theta)}drd\theta \tag{6.11}$$

Applying the projection-slice theorem, this relationship becomes

$$g(\rho\cos(\Theta), \rho\sin(\Theta)) = \frac{1}{4\pi^2}\int_{-\pi/2}^{+\pi/2}\int_{-\infty}^{+\infty} P_\theta(r) \cdot |r|e^{+jr\rho\cos(\Theta-\theta)}drd\theta \tag{6.12}$$

Examining the integral over r, it's clear that this is in the form of an inverse Fourier transform. Noting that the relationship $\rho\cos(\Theta - \theta)$ is the argument of the inverse Fourier transform, the above equation can be written as

$$g(\rho\cos(\Theta), \rho\sin(\Theta)) = \frac{1}{2\pi}\int_{-\pi/2}^{+\pi/2} [p_\theta * k](\rho\cos(\Theta-\theta))d\theta \tag{6.13}$$

where $k = F^{-1}[|r|]$ and $*$ is the convolution operator. Reconstruction is accomplished by evaluating convolutions for each of the pulses (for each θ) and then approximating the integral via sums. The convolutions can be evaluated in the Fourier domain so that the convolutions can be evaluated as a series of element-by-element multiplications, rather than as convolutions in the image domain. This is convenient, as we have $P_\theta(r)$ available. This implies that we should use Equation (6.12).

This equation asserts the integral is evaluated over all r, centered at 0. However, from the last section, we know that each pulse is centered at $K_c = 4\pi f_0/c$ and spans only a small window of r. To deal with this, we will shift $P_\theta(r)$ so that it's centered at 0 and apply a decaying window function to $P_\theta(r)$ before subjecting it to an inverse Fourier transform. This will avoid Gibbs phenomena in the data. With these steps in mind, Equation (6.12) can be rewritten as

$$
\begin{aligned}
g(\rho\cos(\Theta), \rho\sin(\Theta)) = \frac{1}{4\pi^2}\int_{-pi/2}^{pi/2} \\
\left[\int_0^{X_2-X_1} P_\theta(r+X_1)|r+X_1|W(r)e^{+jr\rho\cos(\Theta-\theta)}dr\right] \\
\cdot e^{+jX_1\rho\cos(\Theta-\theta)}d\theta
\end{aligned}
\tag{6.14}
$$

where $X_1 = \frac{2}{c}\left(2\pi f_0 - K_r T + \frac{4K_r L}{c}\right)$ and $X_2 = \frac{2}{c}\left(2\pi f_0 + K_r T - \frac{4K_r L}{c}\right)$

This equation requires that the inverse transform be evaluated over a range of ρ, θ, and Θ for $\rho\cos(\Theta - \theta)$. This means that interpolation is required before summing up over θ. Figure 6.16 illustrates the general flow of the algorithm. First, the inverse Fourier

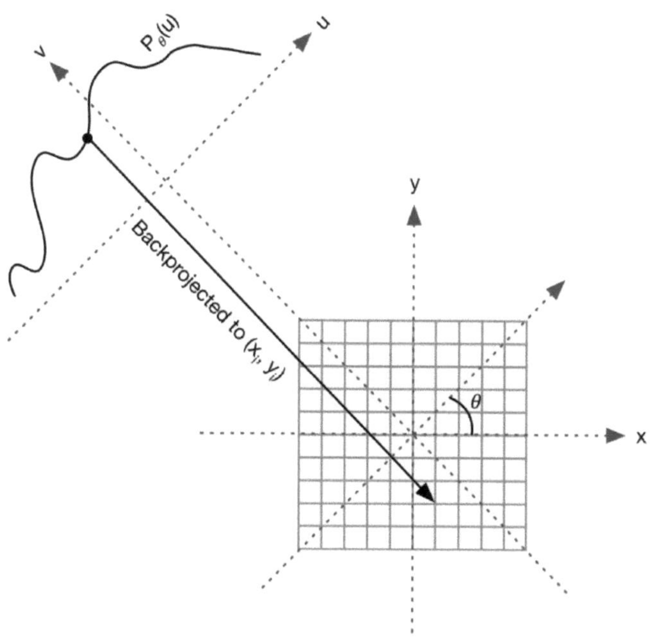

FIGURE 6.16 Illustration of the Backprojection Algorithm (BPA). Each point in the image, represented by the grid, is calculated using the filtered and convolved projection $P_\theta(u)$.

transform of each pulse is taken and then a windowing function is applied, satisfying the $P_\theta(r + X_1)|r + X_1|W(r)$ part of the integral. Then, an r is calculated for each point in what will become the reconstructed image. One-dimensional interpolation is performed for each grid point in the reconstructed image. This is then added to all of the previous θ pulse results. The image will start to focus after some number of pulses are added together. Once all of the pulses have been integrated together, a fully focused image will have been formed. This image should be at the resolution specified by $\Delta\theta$ in azimuth (and by bandwidth in range).

One advantage of BPA is that it is very easy to parallelize. The algorithm can easily be coded for a multiprocessor or graphics processing unit (GPU) and each pulse projected independently. Those indepedent coherent projections are then summed together for the final focused SAR image.

CHAPTER 7
Autofocus

As the synthetic aperture radar (SAR) data is collected, there are many opportunities for phase error to collect. Some of these errors come from approximations made or that not all of the motion effects of the aircraft or spacecraft are well compensated. In these cases, the image can be significantly degraded, which takes the visual form of "defocusing" the imagery. That is, the imagery looks smeared across azimuth (akin to camera shake or smear). Autofocus is an area of study that attempts to approximate the phase errors across the aperture and remove them. If successful, the image quality can be improved significantly, removing the smear and restoring detail to the image. Two techniques will be presented, Map Drift (MD) and Phase Gradient Autofocus (PGA). Some of these techniques are used in other coherent imaging modalities, such as some types of medical imaging.

7.1 Sources of Phase Error

In building the mathematical models associated with SAR, we have made several qualifying assumptions. These assumptions include:

- The wavefront curvature be essentially linear across the ground patch being illuminated,
- The ground patch be flat,
- The ground patch be stationary,
- For dechirp processing, the residual phase term be sufficiently small,
- And, finally, the precise position of the sensor platform be known for the entire dwell.

With the exception of the second assumption, if any of the other assumptions are violated, then phase errors will occur. The assumption that the ground patch be flat will not generate phase errors, but will create a phenomenon in the SAR imagery called *layover*, which will be illustrated later. The most common phase error is called *quadratic phase error* (QPE). Phase errors associated with a moving object are localized to the moving object itself. As will be discussed in Chapter 12, moving objects will shift in image space (based on their velocity in the range direction) and smear in the azimuth direction (based on their velocity in the azimuth direction). This creates a localized defocusing, which can either be used in automatic target recognition to identify a moving object, or these phase errors can be removed using a different set of autofocus algorithms to focus the object for visual or automatic identification.

As we develop the error model, it will be shown that phase errors induced in a single pulse will end up being linear and will cause a modest, often negligible, shift in range from pulse-to-pulse. However, phase errors across the aperture often can be substantial. These phase errors are typically due to inaccuracies in estimating the position and motion of the moving platform upon which the sensor is deployed. For truly accurate phase estimation, the inaccuracies in phase need to be on the order of a fraction of a wavelength. For an X-band system, this means estimating the position in range within a fraction of 3 centimeters. This becomes an incredibly difficult task, particularly aboard aircraft platforms. As apertures grow in size and duration, the estimation of the position and kinematics of the aircraft can drift.

Recall in Chapter 4, the position and kinematics of an aircraft are typically measured by a GPS/inertial measurement unit (IMU) or an inertial navigation system (INS). These systems typically use motion filters, such as a Kalman filter, to anticipate and estimate changes in position and kinematics over time. When the motion is smooth, the system often relies on this estimation, rather than direct measurement. As an imaging time increases, the changes of drift in these position/kinematic estimation increase. Additionally, perturbations in the motion that are not well-behaved (e.g., nonlinear motion) will often cause inaccuracies in estimation of position, velocity, and acceleration. As pulses are transmitted, received, and decoded, the SAR will also record the estimations of the position (in space), orientation of the craft (roll, pitch, yaw), and kinematics (velocity and acceleration). As we build the error model across the aperture, it will become clear that the most prominent type of phase error is QPE, with higher-order phase errors (HOPEs) also a concern.

Autofocus algorithms concern themselves with estimating the phase error across the aperture and removing those phase errors to create a high-quality magnitude image. As will be shown, significant amounts of phase error across the aperture will significantly degrade the quality of the SAR image, in some cases making the image unusable either for human or machine interpretation. Autofocus restores the image quality, often dramatically, for visual or automatic inspection. Two algorithms will be discussed here: Map Drift and Phase Gradient Autofocus. MD requires that the order of the error be well understood and only polynomial phase error of low order can be estimated. PGA estimates the phase error and makes no assumptions about the behavior of the phase error across the aperture. The phase error can be linear, nonlinear, or random in nature. PGA is used extensively in other coherent imaging systems, such as medical imaging. It uses a maximum likelihood estimation (MLE) technique to estimate the error across the aperture and remove it.

7.1.1 Phase Error within a Pulse

Phase error within a pulse occurs when an assumption about the demodulation has occurred. In the case of quadrature demodulation, an error in the matched filtering will create a phase error (e.g., a shift in range because the range to mid-swath was not well known). In the case of dechirp processing, an error in the reference chirp will create a phase error. As we build the mathematical model in this section, we will see that this error ends up being linear. Linear phase error translates to a shift in range in the range-compressed data. In most cases, the linear phase errors end up being less than a resolution cell, meaning that the error is negligible.

For the sake of simplicity, let us start with a single pulse that has undergone dechirp processing. As we saw in Chapter 6, this can be modeled as

$$r(t) = G(U) = \int_{-u_1}^{+u_1} p_\theta(u)e^{-juU}du \tag{7.1}$$

where U is defined as

$$U = \frac{2}{c}\left(2\pi f_0 + \pi K_r(t - \tau_0)\right) \tag{7.2}$$

where t is fast time, τ_0 is the round-trip time delay to the scene center, f_0 is the center frequency of the radar, and K_r is the chirp rate of the radar. These are the spatial frequencies that correspond to each range bin. This form of $r(t)$ is valid only when the residual video phase term is ignored. Recall that $p_\theta(u)$ is the projection of the target reflectivity density function $g(x, y)$ at a given squint angle θ.

Recall from Chapter 2 that each pulse is a time-delayed and attenuated version of the original transmitted signal. Equation (2.5) shows the mathematical form of this. Ignoring the attenuation and windowing terms, this reduces to

$$s_r(t) = \cos\left(2\pi f_0(t - \tau) + \pi K(t - \tau)^2\right) \tag{7.3}$$

For dechirp processing, a reference signal is mixed with the receive signal to generate a single sinusoid at a given range. In Chapter 2 we assumed that the reference signal was perfectly referenced to the scene center by τ_c. Let us now assume that there is some error in the reference signal ϵ. Now the reference signal becomes

$$\cos\left(2\pi f_0(t - \tau_c + \epsilon) + \pi K(t - \tau_c + \epsilon)^2\right) \tag{7.4}$$

for the I channel and

$$\sin\left(2\pi f_0(t - \tau_c + \epsilon) + \pi K(t - \tau_c + \epsilon)^2\right) \tag{7.5}$$

for the Q channel. After dechirp processing, the output becomes

$$\begin{aligned} r_\epsilon(t) = G_\epsilon(U) &= \int_{-u_1}^{+u_1} p_\theta(u)e^{-j\pi\epsilon^2 K_r}e^{+j\frac{\epsilon c}{2}U}e^{-juU} \\ &= e^{-j\pi\epsilon^2 K_r}e^{+j\frac{\epsilon c}{2}U}\int_{-u_1}^{+u_1} p_\theta(u)e^{-juU} \end{aligned} \tag{7.6}$$

This ignores the attenuation and window terms. Consider the possibility that ϵ is well-behaved and less than a resolution cell. This implies, mathematically, that

$$\frac{\epsilon c}{2} << \frac{c}{2B} \tag{7.7}$$

where the left-hand side is the contribution of phase error in range in unit meters, while the right-hand side corresponds directly to range resolution. So, if we have a SAR system that operates at a nominal range resolution of 1 meter, then ϵ has to be much smaller than the reciprocal bandwidth of the system, which would be 150 MHz. That is,

$$\epsilon << \frac{1}{B} \tag{7.8}$$

or 6.67×10^{-9} for this hypothetical case. Coupling the relationship above with the chirp rate, we get

$$\epsilon^2 K_r << \frac{1}{T_p B} \tag{7.9}$$

where the right-hand side is the time-bandwidth product. Most high-resolution SAR systems have very large time-bandwidth products. This implies that the first term in Equation (7.6) is approximately unity, so the equation simplifies to

$$r_\epsilon(t) = G_\epsilon(U) = e^{+j\frac{\epsilon c}{2}U} \int_{-u_1}^{+u_1} p_\theta(u) e^{-juU} \tag{7.10}$$

It is clear that $r_\epsilon(t) = e^{+j\frac{\epsilon c}{2}U} r(t)$. The $e^{+j\frac{\epsilon c}{2}U}$ is a linear phase in terms of U. This linear phase term adds a constant phase term plus shifts all of the ranges in the pulses by

$$\Delta R = \frac{\epsilon c}{2} \tag{7.11}$$

Again, provided that ϵ is much smaller than the transmitted bandwidth, this implies that ΔR will be small, often much smaller than a fraction of the wavelength of the radar. This means that the shift remains within a single resolution cell and is negligible at the radar's defined range resolution.

Providing that position can be estimated to within a fraction of a wavelength for a high-resolution SAR system, we can safely ignore errors in demodulation.

7.1.2 Phase Error across the Aperture

Let us now concern ourselves with errors across the synthetic aperture. In the previous section, we assumed a single error term ϵ that could be safely ignored within a single pulse. Over the course of a dwell, each pulse may well have different ϵ values. This means that, during imaging, phase error should be considered across the synthetic aperture $\epsilon(\eta)$. As we saw, the phase error function is constant in range. But in azimuth, the aperture phase error may vary arbitrarily. Since that phase error in range is constant, the phase error across the aperture is induced along range bins. We can write this as

$$G_\epsilon(\eta, t) = e^{+j\phi(\eta)} G(\eta, t) \tag{7.12}$$

where, in this case, G is the range-compressed data (but not compressed in Doppler), $\phi(\eta)$ defines the phase error across the aperture and $\phi(\eta) = 2\pi f_0 \epsilon(\eta)$. Not surprisingly, since range is approximately parabolic over slow time, even a constant ϵ will induce a parabolic phase error. This parabolic phase error is, in fact, QPE.

Equation (7.12) defines how phase in the phase history domain is modified by the error term over the aperture. To get to the image domain, an inverse Fast Fourier Transform (FFT) is performed on each range bin in the range-compressed domain,

$$\begin{aligned} g_\epsilon &= \text{IFFT}_t[G_\epsilon(\eta, t)] \\ &= \text{IFFT}_t[e^{+j\phi(\eta)}G(\eta, t)] \\ &= \text{IFFT}_t[e^{+j\phi(\eta)}] * g(\eta, t) \end{aligned} \tag{7.13}$$

where $g(\eta, t)$ is the uncorrupted SAR image and "$*$" denotes the convolution operator. This implies that the phase error $\phi(\eta)$ is convolved with the uncorrupted image, $g(\eta, t)$ to create a defocused image. The extent of the defocus is directly related to the magnitude of $\phi(\eta)$, as well as its form.

The most common error in estimating the kinematics of the moving platform is estimating velocity. Figure 7.1 illustrates range over the aperture, separated by some distance d from the ground reference point (GRP). The range affected by velocity is

$$R(\eta) = \sqrt{R_0^2 + (v_a(\eta)t)^2} \tag{7.14}$$

where v_a is the velocity of the aircraft. We are assuming that it is changing over some time in the synthetic aperture. We can make the following assertion regarding the

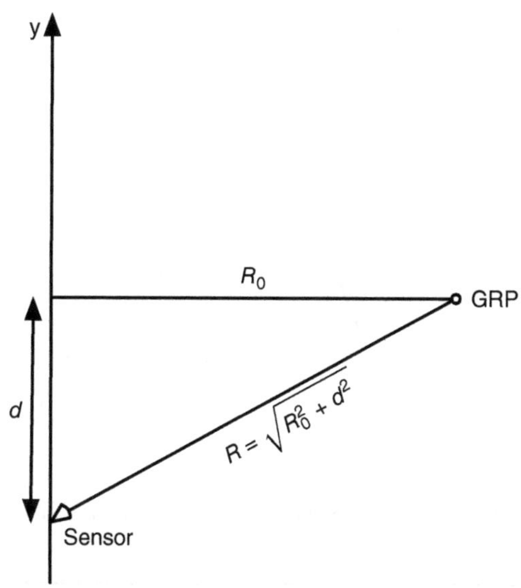

FIGURE 7.1 Simple geometrical model illustrating range over slow time.

demodulation time in the reference signal,

$$\begin{aligned}\tau_c(\eta) &= \frac{2}{c}\sqrt{R_0^2 + d^2} \\ &\approx \frac{1}{c}\left(2R_0 + \frac{d^2}{R_0}\right) \\ &\approx \frac{1}{c}\left(2R_0 + \frac{(v_a\eta)^2}{R_0}\right)\end{aligned} \tag{7.15}$$

If we consider v_a to be the correct velocity and v_e to be the estimated velocity, then the error in timing can be shown to be

$$\epsilon(\eta) = \frac{1}{cR_0}\left(v_a^2 - v_e^2\right)\eta^2 \tag{7.16}$$

As mentioned previously, this results in a QPE.

Example 7.1 Consider an X-band SAR system, flying at 150 m/s with a center frequency of 10 GHz. The range from the SAR to the scene center at the point of closest approach is 20 kilometers. For a hypothetical collection, the system's INS unit has an error of −2 m/s when estimating velocity. That is, the aircraft is flying at 150 m/s, the system estimates the velocity to be 148 m/s. This erroneous estimation leads to a timing error over slow time on the order of a tenth of a nanosecond at either end of the collection. Phase error can be calculated by

$$\phi(\eta) = 2\pi f_0 \epsilon(\eta) \tag{7.17}$$

Figure 7.2 illustrates the quadratic timing error and quadratic phase error when the hypothetical system is off by 2 m/s in its velocity estimation. Given this, the estimated phase error is as much as 350° at the end of the collection. That is, the phase is out-of-sync by as much as 350° at the start and ends of the collection, with the error

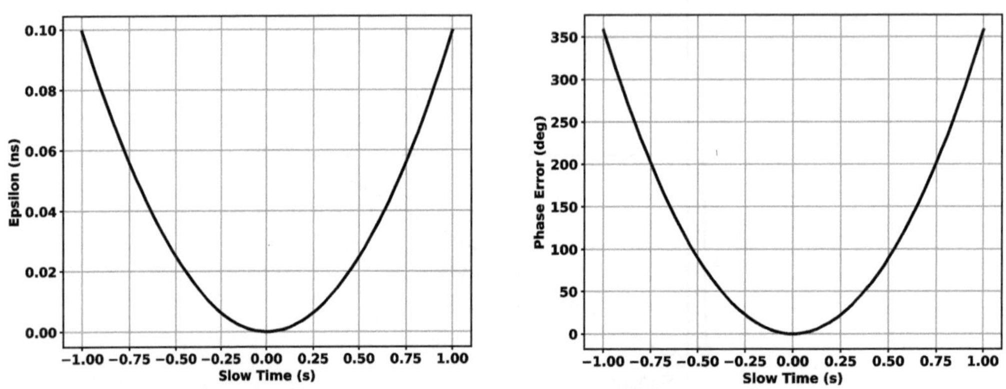

FIGURE 7.2 The quadratic timing error (on the left) and the quadratic phase error (on the right) when the hypothetical system described in the example is off by 2 m/s in its velocity estimation.

in the center of the dwell to be zero. If not corrected, the resulting phase error will defocus the image, smearing across azimuth.

It can be shown[1] that the maximum velocity error that can be tolerated before image degradation occurs is

$$\Delta v_{\max} = \frac{v_a}{\lambda R_0} \rho_a^2 \tag{7.18}$$

where v_a is again the actual velocity of the aircraft, λ is the center wavelength of the radar, and ρ_a is the azimuth resolution. For this example, the maximum error that can be tolerated before smearing occurs is 0.25 m/s. There is a direct correlation between azimuth resolution and maximum velocity estimation error. The coarser the resolution, the more forgiving the maximum velocity estimation error.

Poorly estimated velocity is a common error, but it's not the only source of error. HOPEs can occur when there are other types of uncompensated motion effects or timing errors due to propagation through the atmosphere. Equation (7.13) indicates that if we can estimate the phase error across the aperture, then we can compensate for it in the range-Doppler space and then restore image quality. The estimation of phase error will be discussed in the next two sections. The estimation of phase errors and the compensation of those errors comprise the class of algorithms known as *autofocus* algorithms. Both of the algorithms presented here estimate that phase error directly from the SAR data.

MD estimates the phase error through a technique called subaperturing. Subaperturing divides the pulses up into equal divisions, or subapertures, and then compares the data within those subapertures to each other to estimate phase error. The estimated phase error is removed and the image quality is restored. As will be discussed, the MD algorithm requires some assumptions on the order of the phase error. MD cannot remove nonlinear or random phase errors from the aperture.

PGA uses the gradient of phase across the entire aperture to estimate the phase error. This is done with a maximum likelihood estimator (MLE). The advantage to this technique is that the form of the error does not need to be defined. Any arbitrary phase error function can be estimated and compensated. PGA is used extensively in SAR, as well as in medical imaging.

7.2 Map Drift

The technique of *map drift*[2] estimates the phase error across the aperture by employing *subaperturing*. As we'll see in Chapter 8, SAR data is holographic. That is, only a portion of the phase is needed to faithfully reconstruct the image. There were some hints of this in the summary of the Backprojection Algorithm (BPA).

MD assumes that the aperture phase error takes the form of a finite polynomial expansion. That is, the phase can be fully described by either a quadratic function or a cubic function. The technique leverages the subapertures to estimate the parameters in

[1] C. V. Jakowatz, D. E. Wahl, P. H. Eichel, D. C. Ghiglia, and P. A. Thompson, *Spotlight-Mode Synthetic Aperture Radar: A Signal Processing Approach*, Kluwer Academic Publishers, Boston, MA, 1996.

[2] T. M. Calloway and G. W. Donohoe, "Subaperture Autofocus for Synthetic Aperture Radar," *IEEE Transactions on Aerospace and Electronic Systems*, Vol. 30, No. 2, pp. 617–621, 1994.

the polynomic function and does so with all of the data across the aperture, split into subapertures according to the order of the polynomial required. That is, if the order of the polynomial is quadratic, then only two subapertures are needed. Cubic functions require three subapertures, and so on.

To describe the algorithm, we will assume a QPE function. That function will be modeled as

$$\Phi(m) = bm^2 \tag{7.19}$$

where m is the aperture position index and b is the unknown coefficient that must be estimated. The unknown coefficient b is estimated from two subapertures of the image $s_{\text{dc}}(\eta, t)$. The phase data is split into two even chunks and a 2D FFT is performed to transform the image from the phase history domain to the image domain. The algorithm relies on the idea that the two phase histories are essentially the same. Their azimuth resolutions will have been reduced, but the two scenes will be similar.

The algorithm flow for MD is shown in Figure 7.3. The algorithm starts by taking the complex data in the phase history domain and chunking it into two even chunks in slow time. Each chunk is a subaperture of the total aperture. A 2D FFT would put the data back into the image domain and a comparison of each subaperture would show two seemingly identical images. However, each image is formed from a different viewing geometry than the other. Additionally, they are formed by splitting the aperture into two, where one side of the quadratic is in one image and the other side of the image is in the other. That is, the resulting subapertures are affected by half of the original QPE. The "half" of the QPE can be estimated as a linear. This can be seen in Figure 7.4.

Once the each subaperture has been transformed back to the image domain, the linear component of the quadratic in the subaperture creates a translational shift due to the shift property of the Fourier transform. To estimate this shift, we can turn to cross-correlation or template matching. Let $f = |S_1|$, where $|S_1|$ is the magnitude of the first subaperture in the image domain. Then, $g = |S_2|$. Let F and G be the 2D Fourier transforms of the f and g, respectively. The cross-correlation between f and g can be defined as

$$c = \mathcal{F}^{-1}\left\{F^* \cdot G\right\} \tag{7.20}$$

where $\mathcal{F}^{-1}$ is the inverse Fourier transform, F^* represents the complex conjugate of F, and $\cdot$ represents element-by-element multiplication. The linear shift can be measured by

$$\sigma = \arg\max\left\{|c_{\text{mid}}|\right\} \tag{7.21}$$

where c_{mid} is the range bin in the middle of the correlation image. The peak QPE is equivalent to

$$\text{QPE}_{\text{peak}} = \pi\sigma \tag{7.22}$$

Thus, a phase correction can be applied to each range bin that takes the form of

$$\phi(m) = \text{QPE}_{\text{peak}} m^2 \tag{7.23}$$

and is multiplied to each range bin in the phase history domain by $e^{-j\phi(m)}$.

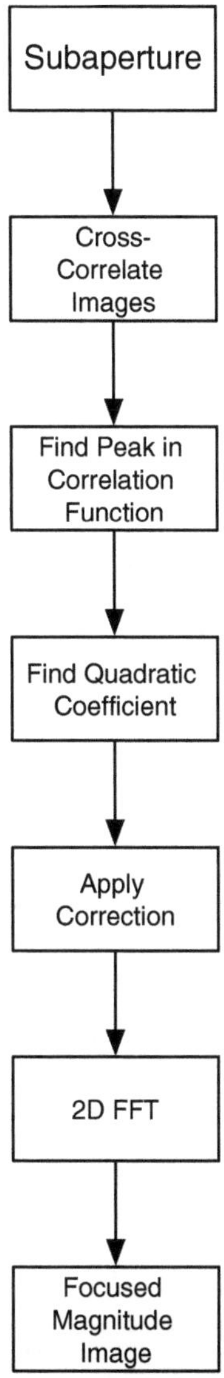

Figure 7.3 The flow for the Map Drift algorithm.

Example 7.2 Consider the following X-band SAR image shown in Figure 7.5, taken by Sandia National Laboratories as part of the Defense Advanced Research Project Agency (DARPA) program called "MSTAR." The data, taken in the mid-1990s, is of a rural area outside of Huntsville, AL. At the time of this writing, the image set is freely available from Air Force Research Laboratories (AFRL).[3]

The image is well focused, showing a road, trees, and fields. Quadratic phase error can be added to the image, causing it to defocus. A peak quadratic phase error of 30 radians (or about 1719°) is introduced. The quadratic phase error function and

[3] https://www.sdms.afrl.af.mil/index.php?collection=mstar

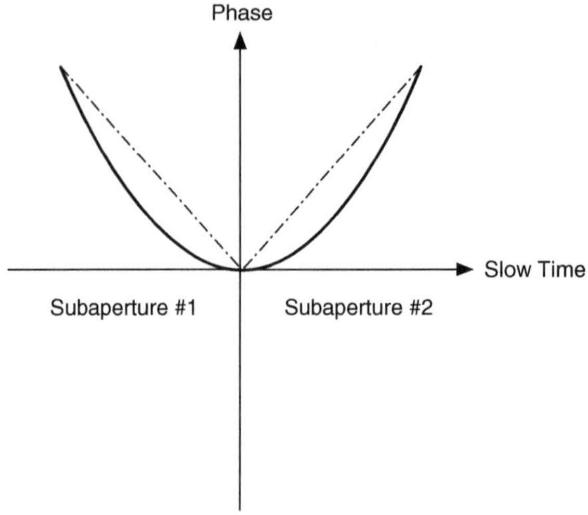

Figure 7.4 The modeled quadratic phase error, split into two subapertures, with the approximate linear components for each subaperture.

the defocused image are shown in Figure 7.6. The smearing in the azimuth direction is visually noticeable.

Two subapertures were created by taking the defocused image back into the phase history domain using an inverse 2D FFT and splitting that data into two in the slow-time dimension. Another set of 2D FFTs puts the data back into the image domain. The two subapertures are shown in Figure 7.7.

The linear shift between the two subapertures is not visually discernable. The cross-correlation step is necessary to measure this shift in pixel space. The cross-correlation is shown in Figure 7.8. The peak is plainly visible in the 2D

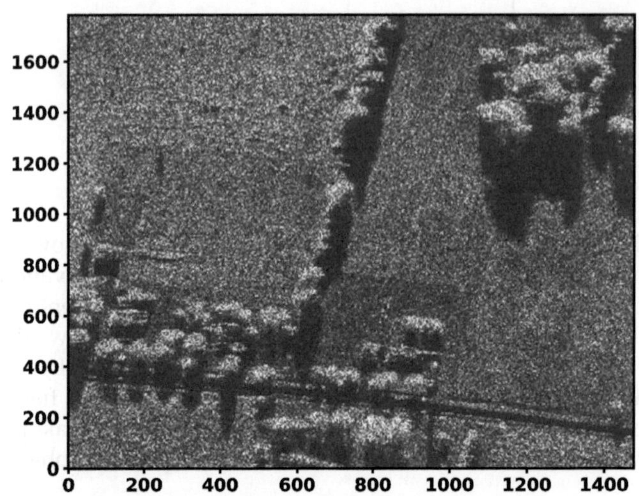

Figure 7.5 An example image from the MSTAR clutter data set.

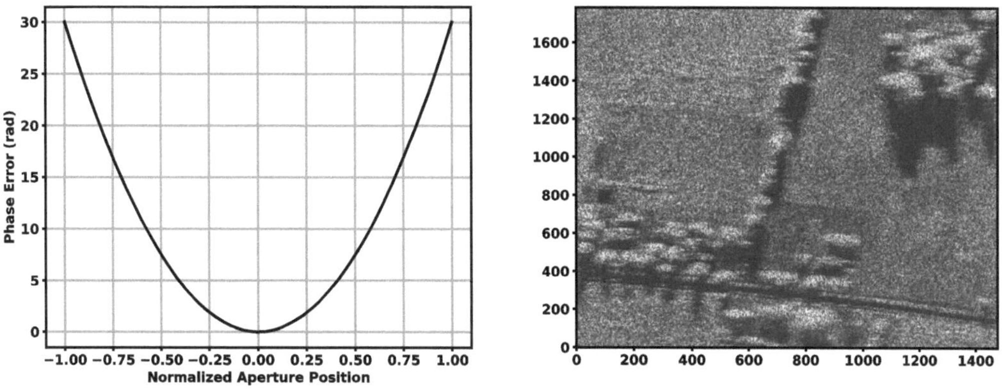

Figure 7.6 MSTAR image with 30 radians of quadratic phase error added.

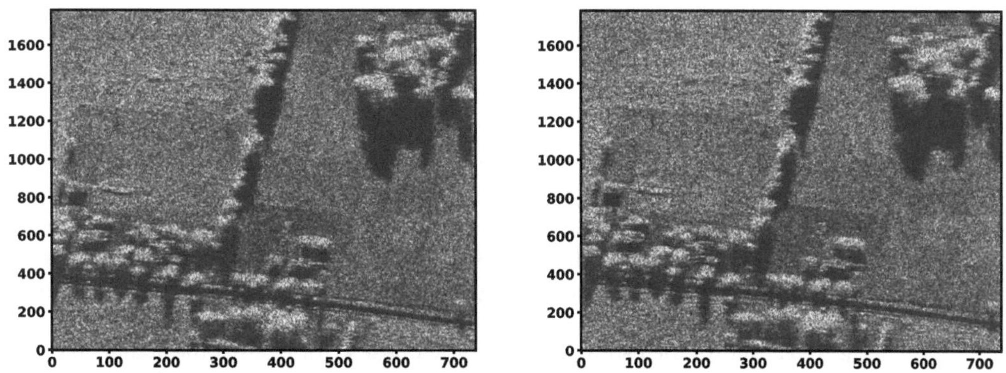

Figure 7.7 Two subapertures created from the phase history of the defocused MSTAR image. The linear shift between the two subapertures is not visually discernable.

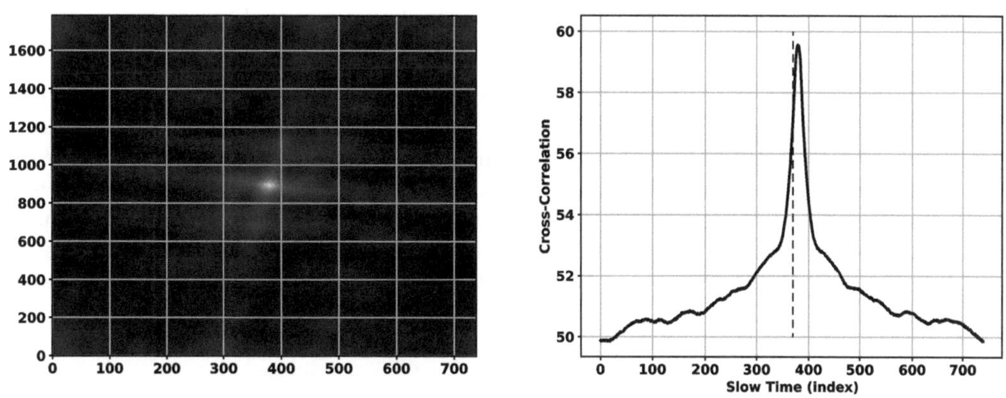

Figure 7.8 The 2D cross-correlation image on the left, with a plot of the center range bin with the correlation peak on the right. The shift off the center of the azimuth represents σ.

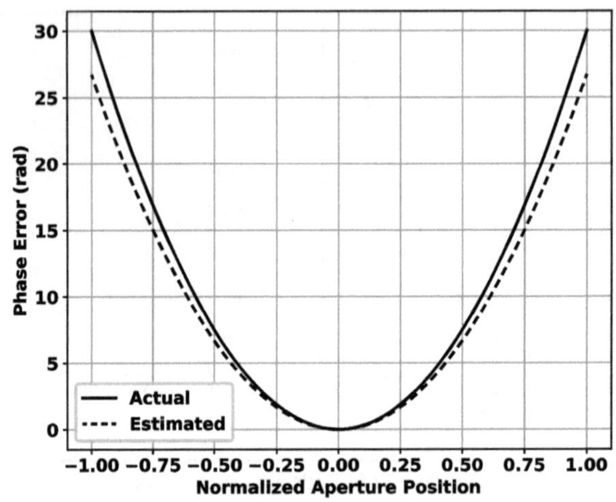

FIGURE 7.9 The estimated quadratic phase error from the Map Drift (MD) algorithm versus the actual induced quadratic phase error.

cross-correlation image, shown on the left of the figure. Plotting the center range bin, the peak is also plainly visible, shown on the right of the figure. The center of the plot is shown by the dotted line. The distance, in pixels, between the peak and that center is σ. Thus, the peak quadratic error is $\pi\sigma$. In this case, the peak is estimated to be about 8.5 pixels from the center, suggesting that the estimated QPE is about 26.7 radians.

Figure 7.9 shows the estimated quadratic phase error against the actual 30 radians of quadratic phase error that was injected into the phase history. The estimated is about 10% from actual. Figure 7.10 shows the image after the phase correction is applied to the full phase history (not the subapertures) and a 2D FFT puts the data back into the image domain. While the corrected image has some residual phase error across the aperture, it is not enough to sacrifice image quality.

Calloway and Donohoe (1994)[4] imply in their paper that this technique can be extended to higher polynomial forms. Each requires another subaperture and then the cross-correlation of the subaperture with the polynomic form provides an estimate of a coefficient to the model. Thus, since the cubic form has two coefficients, then three subapertures are needed. A 4th order polynomial has three coefficients, so four subapertures are needed. There is a point of diminishing return in creating *N* subapertures, but most phase errors can be corrected using a relatively low-order polynomial model.

[4] T. M. Calloway and G. W. Donohoe, "Subaperture Autofocus for Synthetic Aperture Radar," *IEEE Transactions on Aerospace and Electronic Systems*, Vol. 30, No. 2, pp. 617–621, 1994.

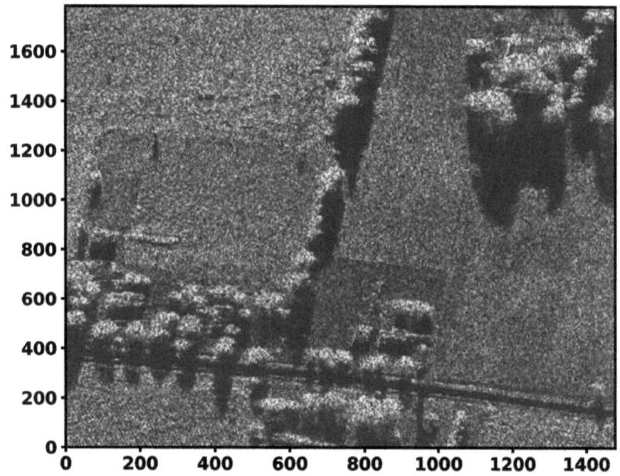

FIGURE 7.10 The focused image after Map Drift (MD) is applied.

7.3 Phase Gradient Autofocus

While the results of the autofocus for MD appears to be successful and relatively fast, the technique is parametric, requiring that the user known, or at least assume, the order of the phase error across the aperture. There are many cases when just removing QPE will be acceptable, but there are other sources of phase error, including turbulence and atmospheric interference. These are higher-order phase errors and, in some cases, nonlinear. A technique that can estimate and remove these errors is desirable, particularly if the technique is automatic, requiring little to no human-in-the-loop interaction.

PGA[5] fits this bill nicely, being data driven, nonparametric, and fast. The algorithm leverages the notion that every scatterer is, in fact, influenced by the phase error induced across the aperture. By using scatterers across the image, the algorithm can obtain a more accurate estimate of the phase error. It does this by using maximum likelihood estimation, which is a formal optimal estimation theory.

As Jakowatz et al. (1996) point out,[6] a straightforward way to incorporate the information from multiple targets in the scene would be to estimate the phase error for each target and then average the ensemble of phase errors into one estimate. Jakowatz et al. (1996) point out that for very low signal cases, the phase estimate from any scatterers is essentially a random variable. That random variable is uniformly distributed between $-\pi$ and π. An average of those phases, particularly with stronger scatterers, would not produce a useful estimate of the phase error that could then be removed. In some cases, it may actually make image quality worse.

---

[5] D. E. Wahl, P. H. Eichel, D. C. Ghiglia, and C. V. J. Jakowatz, "Phase gradient Autofocus: A Robust Tool for High Resolution SAR Phase Correction," *IEEE Transactions on Aerospace and Electronic Systems*, Vol. 30, No. 3, pp. 827–835, 1994.

[6] C. V. Jakowatz, D. E. Wahl, P. H. Eichel, D. C. Ghiglia, and P. A. Thompson, *Spotlight-Mode Synthetic Aperture Radar: A Signal Processing Approach*, Kluwer Academic Publishers, Boston, MA, 1996.

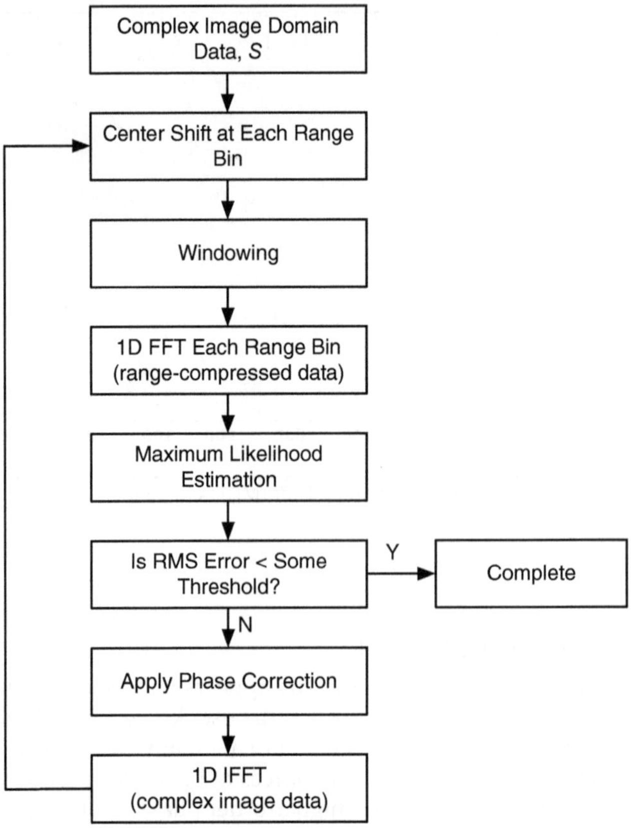

Figure 7.11 The algorithm flow for Phase Gradient Autofocus.

The PGA algorithm instead uses an MLE to iteratively estimate the phase error across the aperture. It has been shown that the MLE is the optimal estimator by comparing the performance of the algorithm to the Cramer-Rao lower bound on estimator variance. It is important to note that the estimator in Wahl et al. (1994) is a linear unbiased minimum variance (LUMV) estimate of the gradient, while the estimator described in Jakowatz et al. (1996) is the MLE. Careful reading will also show that original paper on PGA[7] used a different estimator all together. This chapter will focus on the MLE described in Jakowatz et al. (1996).

The algorithm flow for PGA is shown in Figure 7.11. The basic concept is to locate each of the brightest targets in each range line in a magnitude image of the scene. The targets are then shifted to the center of the scene using a circular shift at each range bin. A window is then applied to the data so that only the bright targets, now at the center of the scene at each range line, are retained and everything else is zeroed out. The complex data

[7] P. H. Eichel, D. C. Ghiglia, and C. V. Jakowatz, Jr., "Speckle Processing Method for Synthetic Aperture Radar Phase Correction," *Optics Letters*, Vol. 14, pp. 1101–1103, January 1989.

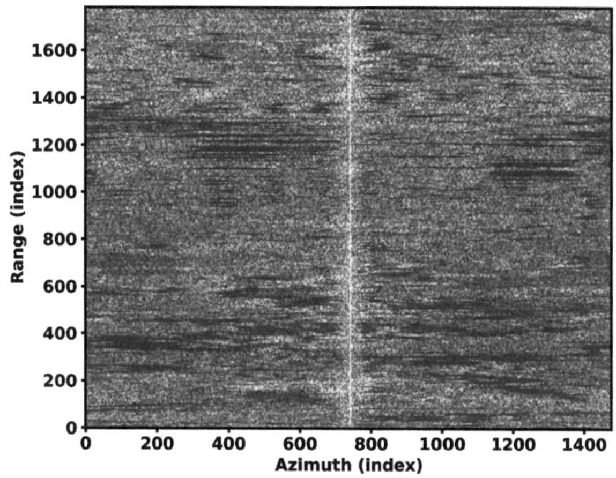

Figure 7.12 The image after the circular shift at each range bin.

is then transformed to the range-compressed domain by performing an FFT at each range bin. Once that's complete, the phase error is estimated by using the phase difference (i.e., the gradient) between two successive pulses. These differences are then integrated to estimate the phase error across the entire aperture. We will walk through the algorithm using the degraded MSTAR image from the previous section, shown in Figure 7.6. Recall that this image had 30 radians of QPE added to it.

7.3.1 Center Shift at Each Range Bin

The first step in the PGA is to take the magnitude data $s_{\text{mag}} = |S|$ and to detect the brightest object in each range bin. That is, the object that has the highest magnitude value is defined to be the brightest. Then, using a circular shift, move that bright object to the center of the image. If done properly, the resulting image will be the same as Figure 7.12. It is the complex data in the image domain that should be circularly shifted.

7.3.2 Windowing

To improve the estimate of the phase gradient, and thus, the phase error, a window is placed over the data at the center of the scene to mask out the clutter and noise. These are not associated with the brightest object in each range line and could corrupt the phase estimate. The width of window is determined by averaging across all of the range bins to produce a single center shift profile. From there, the profile is placed in units of decibels. Jakowatz et al. (1996) recommend that a threshold of −10 dB for a normalized center shift profile. The width is then determined from those values above the −10 dB threshold. Jakowatz et al. (1994) also recommend increasing the window by 50%. Figure 7.13 shows this profile, along with the window that will be applied to the data.

Figure 7.14 shows the data after it's been windowed. The center-shifted targets in each range bin make a sharp line up and down the length of the image.

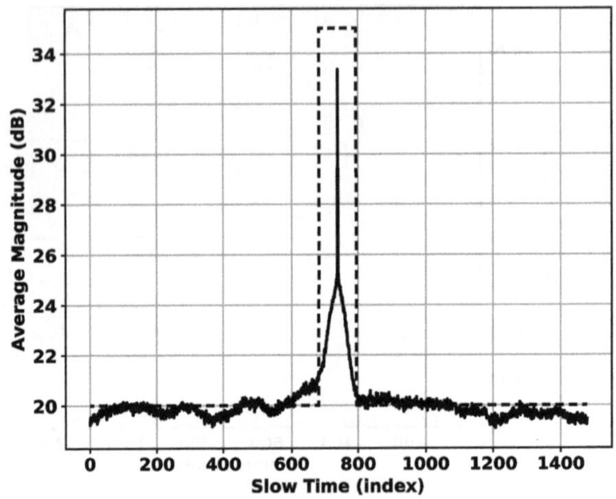

FIGURE 7.13 The center shift profile, calculated by averaging across all of the range bins, along with the window to be applied to the center-shifted data.

7.3.3 Fourier Transform to Range-Compressed Domain

The next step in the algorithm is to put the data in the range-compressed domain, also known as the range-Doppler domain. Since the data currently resides in the range-azimuth domain, a series of FFTs along each range bin will transform the data into the range-Doppler or range-compressed domain. Figure 7.15 shows the data after mapping it from the range-azimuth to the range-Doppler domain.

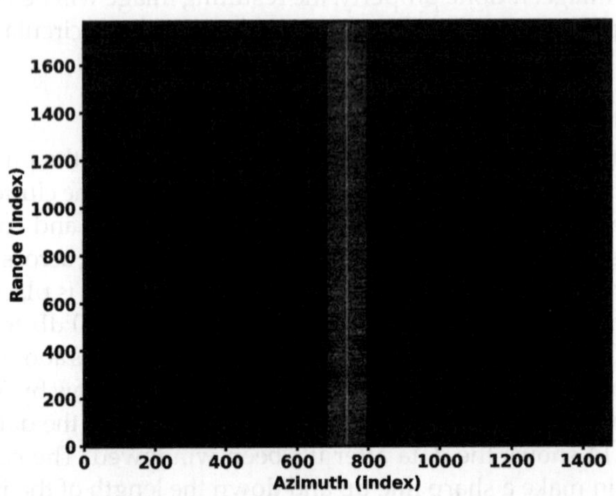

FIGURE 7.14 The center shift complex data after windowing.

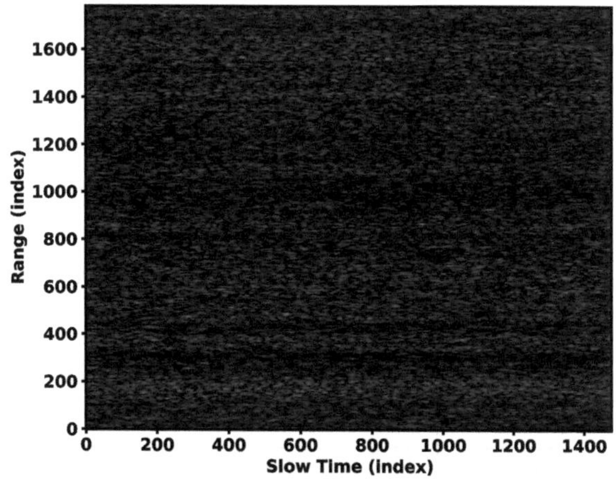

Figure 7.15 Center shift data after FFT at each range bin.

7.3.4 Phase Error Estimation Using the Maximum Likelihood Estimator

The mathematical model for the center-shifted range-Doppler data is

$$g(k,m) = a(k)e^{j\phi(m)} + \eta(k,m) \tag{7.24}$$

where k is the index for samples within a pulse and m is the index for a given pulse, $a(k)$ is the target complex reflectivity, and $\eta(k,m)$ is the noise term. Jakowatz and Wahl (1993)[8] did an extensive analysis to show that the noise term can be assumed to be zero mean, white, Gaussian, and independent of the target complex reflectivity $a(k)$.

The maximum likelihood estimate of $\phi(m)$ uses the phase difference between two adjacent pulses to obtain an estimate for the entire phase error. The phase differences are integrated to obtain this estimate. The phase difference is calculated using

$$\Delta\phi(m) = \phi(m) - \phi(m-1) \tag{7.25}$$

which looks at the difference in phase between adjacent pulses. The maximum likelihood estimator of this phase difference is defined as

$$\widehat{\Delta\phi}(m) = \angle \sum_{k=1}^{N} \{g^*(k,m-1)g(k,m)\} \tag{7.26}$$

[8]C. V. Jakowatz, Jr., and D. E. Wahl, "An Eigenvector Method for Maximum Likelihood Estimation of Phase Errors in SAR Imagery," *Journal of the Optical Society of America A*, Vol. 10, No. 12, pp. 2539–2546, December 1993.

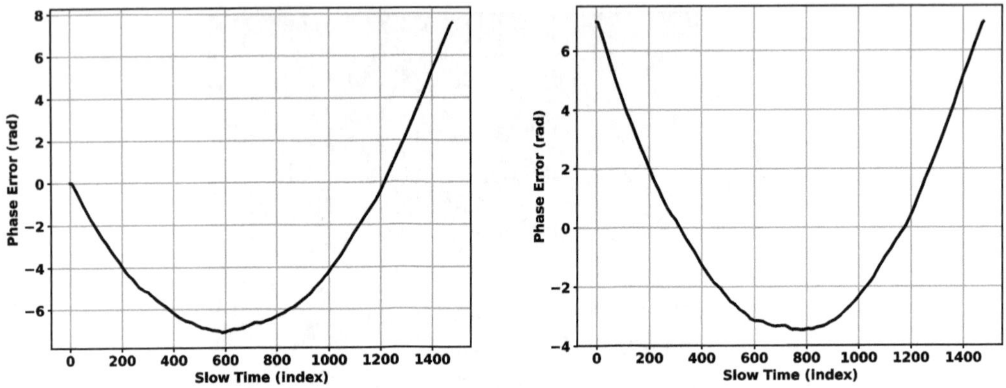

FIGURE 7.16 The entire aperture phase error $\hat{\phi}(m)$. The plot on the left shows the phase error before linear detrending and the plot on the right shows the phase error after linear detrending.

where "$*$" is the complex conjugate operator. Jakowatz et al. (1996) show that this estimator is unbiased and is efficient over varying signal-to-clutter ratios. Signal-to-clutter is the measure used to distinguish how much of the signal for a given scatterer or set of scatterers sets above the clutter level. This measure is typically in units of decibels.

The entire aperture's phase error is then estimated by calculating the cumulative sum for m,

$$\hat{\phi}(m) = \sum_{l=2}^{m} \widehat{\Delta\phi}(l) \tag{7.27}$$

and $\hat{\phi}(1) = 0$. Figure 7.16 shows this estimate over the entire aperture. Note that while the shape is generally quadratic, there is a linear bias on it. This linear bias should be removed by performing linear regression on the data and then subtracting the linear trend from $\hat{\phi}(m)$. The plot on the right of Figure 7.16 shows the phase error after the linear trend is removed. The phase error is fully quadratic and is centered in the aperture.

7.3.5 Removal of Phase Error

The root-mean-square (RMS) error of $\hat{\phi}(m)$ is then calculated and compared to a threshold. If the threshold has not been reached, then the detrended phase error $\hat{\phi}(m)$ should be removed from the original phase history data (not the center-shifted data) along with each range bin. A 2D IFFT puts the corrected data back into the range-azimuth, or image, domain. If the RMS error has reached the threshold or the RMS error starts to increase, then the algorithm can be halted. For our example, the RMS error reached 0.27 radians within nine iterations. If the threshold has not been reached and the RMS error is still decreasing, then the algorithm is performed again on the newly corrected complex image data.

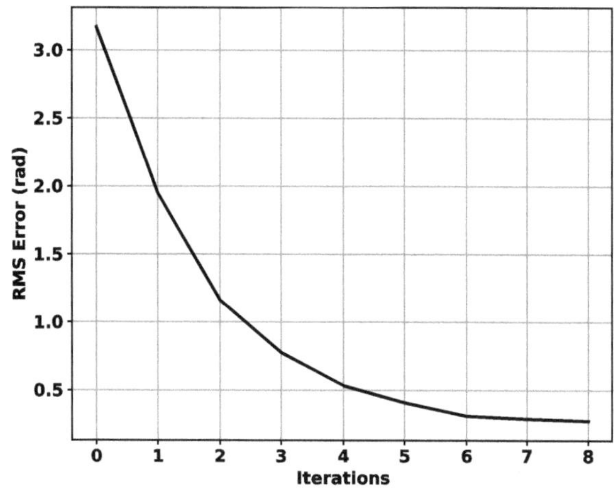

Figure 7.17 The RMS error over each iteration of the PGA algorithm.

Figure 7.17 shows the RMS error over each iteration for our example. The algorithm starts with an RMS error of greater than 3 radians and converges to just under 0.3 radians. The RMS error is a quantitative description of the shape of $\hat{\phi}(m)$. As the shape of the phase error starts to get smaller and flatten, the RMS error gets smaller.

Figure 7.18 shows the final estimated phase error over all of the iterations and the actual phase error induced on the original image. Note that they are very similar despite a nearly 10 radian bias. The difference between the two is shown as a dotted line. The difference is a nearly constant phase term, which provides only a slight shift in

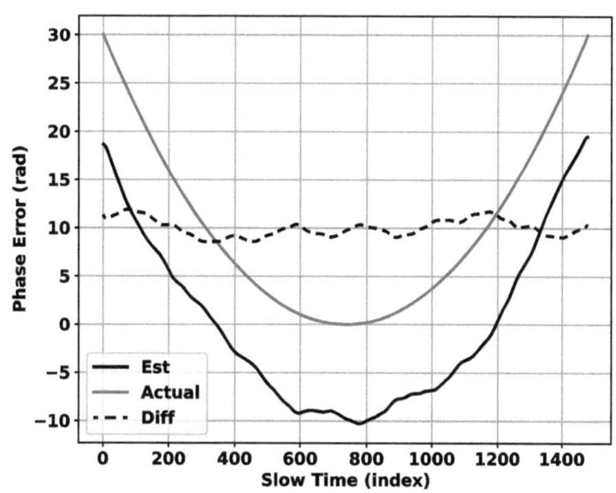

Figure 7.18 The estimated versus actual induced phase error for the MSTAR example.

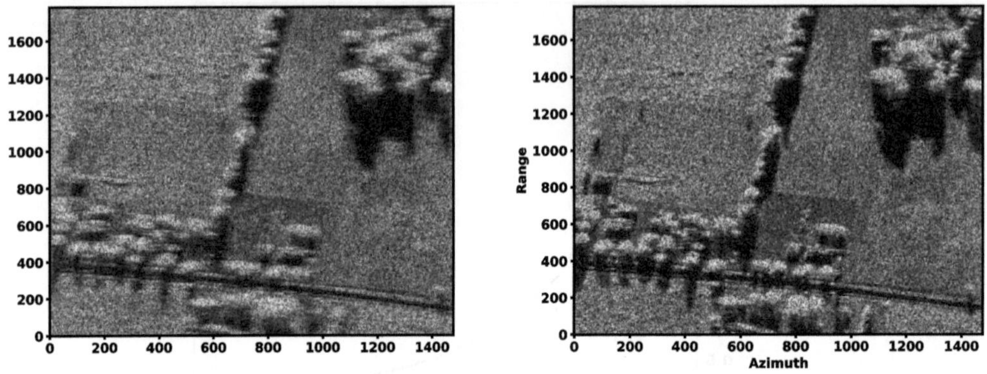

FIGURE 7.19 The defocused and corrected images, side-by-side. The defocused image on the left, while the image corrected with PGA is on the right.

the range dimension. Figure 7.19 shows the before and after for the MSTAR image. The image on the left is the MSTAR image with 30 radians of QPE induced over the aperture. The image on the right is the same MSTAR image after performing PGA, estimating the error, and then correcting the image. The corrected image is well-focused and is a noticeable improvement over the corrupted image.

CHAPTER 8
Image Quality

After image formation is completed, it may be desirable to continue to improve a synthetic aperture radar (SAR) image's interpretability and quality. In particular, speckle noise plays a significant role in the visual quality of a SAR image (a common complaint: SAR is "grainy"). Multilook is a technique by which speckle noise can be significantly reduced, albeit at the expense of loss of resolution. Another undesirable effect is the presence of sidelobes in both range and azimuth; this creates a star-like pattern away from the return for a given scatterer in the scene. Aperture weighting can be used to minimize this effect.

8.1 Speckle

As an active imaging technique, SAR produces its own illumination, and receives the backscatter from that illumination in the form of attenuated and delayed signals from scatterers in the scene. Each pixel is related to a resolution cell and there are likely multiple scatterers within that resolution cell, as the measured returns can be on the order of a fraction of a wavelength. If there are multiple scatterers in a resolution cell (and there likely are), but no one scatterer is stronger than another, then the resolution cell is said to have distributed targets.[1] As we have discussed in previous chapters, this can be represented mathematically as a superposition of returns from the multiple scatterers,

$$Ae^{j\Phi} = \sum_i A_i e^{j\Phi_i} \tag{8.1}$$

where A_i is the complex backscattering coefficient for a given scatterer and Φ_i is the complex phase due to backscatter. These all sum up to be a single attenuation coefficient A and a single phasor Φ. Over each resolution cell, the texture of the scene in that cell is likely to be different than the adjacent cell. Because of this, the attenuation coefficients and phases will appear to be random, and, indeed, follow a random distribution.

[1] F. Argenti, A. Lapini, L. Alparone, and T. Bianchi, "A Tutorial on Speckle Reduction in Synthetic Aperture Radar Images," *IEEE Geoscience and Remote Sensing Magazine*, Vol. 1, No. 3, September 2013.

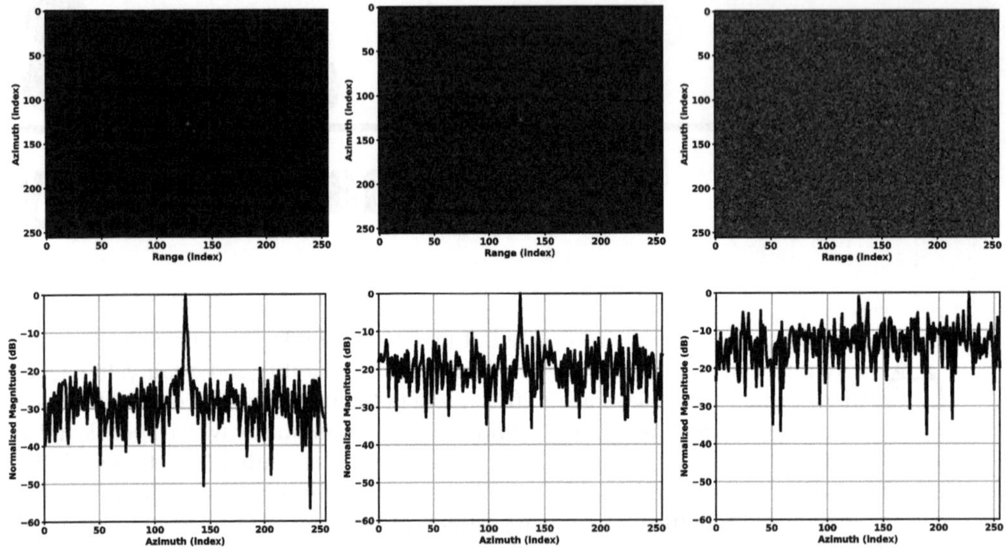

Figure 8.1 Simulation of ideal scatterer embedded in speckle at 30 dB, 20 dB, and 10 dB from left to right, respectively. The second row of images are slices taken at the middle row of the image, which corresponds to the center range bin, and shows the speckle levels changing with respect to the target. The speckle is so destructive at 10 dB that the scatterer is no longer distinguishable from the background.

Figure 8.1 shows speckle at various signal-to-noise ratios (SNRs). Using a linear contrast stretch, where the image is stretch from the minimum magnitude value to the maximum magnitude value, the speckle is hardly noticeable at an SNR of 30 dB. At an SNR of 20 dB, the constructive and destructive nature of the speckle gives the SAR image it's ubiquitous "grainy" look. This granular feature is present in all coherent imaging modalities, from SAR to computer-aided tomography.

The mathematical model for speckle indicates that speckle is a multiplicative noise term. Both A and Φ can be considered randomly distributed and the phases are highly varying. This is because the wavelength is smaller than the resolution cell (usually much smaller) and may be considered uniformly distributed between $(-\pi, \pi)$. The phases are also independent of the attenuation coefficients. If the image is large enough, the central limit theorem applies. This means that the resulting return from a given resolution cell can be seen as a complex signal, whose real and imaginary parts (in-phase and quadrature components) are independent and identically distributed (IID) zero-mean Gaussian.[2,3] This is useful when trying to simulate speckle in an idealized SAR image: speckle can be produced by generating an $N(0, \sigma)$ random sequence for both the I and Q

[2] J. W. Goodman, "Some Fundamental Properties of Speckle," *Journal of the Optical Society of America*, Vol. 66, No. 11, November 1976.

[3] M. Tur, K. C. Chin, and J. W. Goodman, "When Is Speckle Multiplicative?," *Applied Optics*, Vol. 21, No. 7, April 1982.

channels independently, then adding them to the idealized SAR image. It can be shown that σ is derived by

$$\sigma = \frac{\sqrt{(\mu_s - \mu_n)^2}}{4 \cdot \text{SNR}} \tag{8.2}$$

where μ_s is the mean of the idealized SAR image and μ_n is the mean of the Gaussian noise, which should be equal to zero. SNR is the signal-to-noise ratio (NOT in decibels).

With $Ae^{j\Phi} = z_1 + jz_2$, the joint probability density function for z_1, z_2 is

$$p_{z_1,z_2} = \frac{1}{\pi\sigma} e^{-z_1^2 + z_2^2/\sigma} \tag{8.3}$$

The amplitude (magnitude) of A follows a Rayleigh distribution,

$$p_A(A) = \frac{2A}{\sigma} e^{-A^2/\sigma} \tag{8.4}$$

The intensity (or power) of A, $I = A^2$, follows an exponential pdf,

$$p_I(I) = \frac{1}{\sigma} e^{-1/\sigma} \tag{8.5}$$

The mean of the intensity is equal to σ. It can be shown that σ is the radar cross section (RCS) of a given resolution cell. Thus, the received signal can be shown to follow a probability density function of

$$p_{I|\sigma}(I|\sigma) = \frac{1}{\sigma} e^{-1/\sigma} \tag{8.6}$$

or $I = \sigma u$, where u is exponentially distributed. This is the multiplicative model of speckle.

Since we see with noncoherent image systems (our eyes and brain), the granular nature of speckle is often a bane to the interpretability of SAR. It is often the case that individuals who attempt to make use of a SAR image, either by interpreting its contents manually or measuring some portion of the image, complain that speckle is too distracting. This is how SAR got the reputation of being "grainy" and "blobby." So, for visual display and interpretation, it's desirable to minimize the effects of speckle without damaging the other strong scatterer information in the scene. The next couple of sections explore ways in which this can be accomplished.

8.2 Multilook

Because speckle is random and strong scatterers are deterministic, it may be possible to take multiple images of the same scene and incoherently average out the speckle while keeping the strong scatterers untouched. Such a technique is called *multilook*. It can be done with multiple looks of the same scene (although other processing, like registration, may be needed to perfectly align each frame of the multilook). While the resolution of each image remains constant, the search area during a collection is reduced by the number of looks needed to reduce the speckle. So, if a nine-look multilook is desired, eight of the nine images could have been used to search other areas in the scene, instead of looking at the same spot in the scene. Multilook is easier to do with a spotlight imaging mode configuration than with stripmap. However, if only one image is available or search area is of critical importance, then multilook can be accomplished using a technique called *subaperturing*.

8.2.1 Subaperturing

Before we can discuss use of multilook to reduce speckle noise, it's important that we understand the holographic nature of SAR. As mentioned in previous chapters, SAR data is *holographic*. In short, this means that all of the information needed to reconstruct a faithful version of the scene is stored in the phase history and that a portion of the phase history can be used to reconstruct the scene, just at a lower resolution depending on how much of the aperture has been cropped. That is, the phase history or Fourier space acts like a hologram: any selected sub aperture is sufficient to generate a full image whose resolution is dependent on the dimensions of the specified subaperture.

Example 8.1 Consider a scene with a single scatterer at $[-50, -50]$ meters from the scene center. The phase history, also known as the Fourier space, after image formation is a 2D sinusoid. Figure 8.2 shows the real component of the phase history after image formation. Note that the 2D sinusoid is present across most all of the aperture (an inscribed rectangle was used in polar formatting, so the "wings" where there was no specific phase history are noticeable). If the image is $M \times N$ in size, then consider a subaperture that is $M/2 \times N/2$ in size and centered in the image. This is shown in the same figure. However, the placement of this box is somewhat arbitrary; since the 2D sinusoid extends throughout the phase history, any subset of phase history should produce the same image.

Once carved out, a 2D FFT puts the data back into the image domain. Figure 8.3 shows the image with the full aperture (left) and the subapertured image (right).

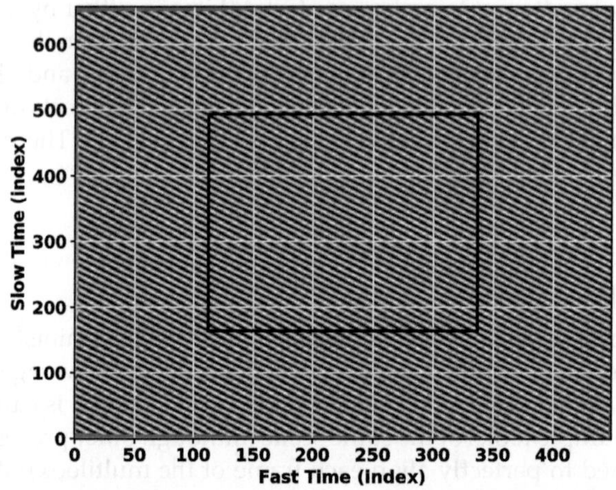

FIGURE 8.2 The phase history of a single target at (−50, 50) meters from the GRP. A subaperture half the size of the full aperture is shown.

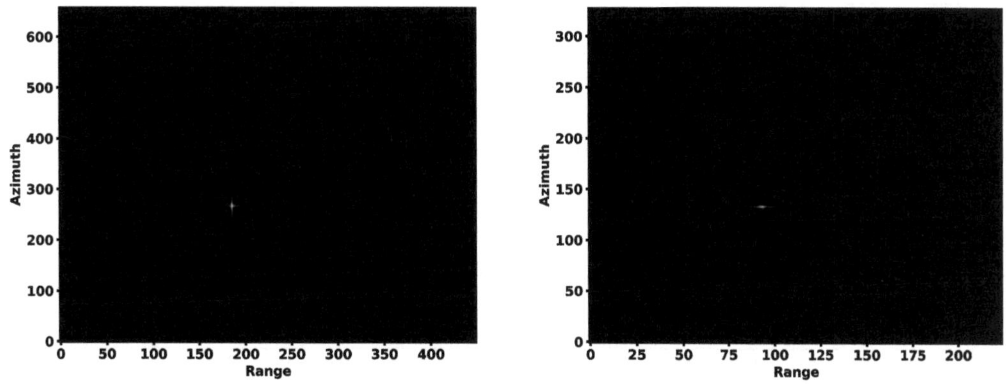

FIGURE 8.3 The results from subaperturing. The image on the left was generated using the full aperture, while the image on the right was generated using the subaperture shown in Figure 8.2.

Note that the subaperture image is half the size in both the range and azimuth direction (a quarter of the entire aperture). While this example exhibits symmetry in the subaperturing (half in both directions), this is not a requirement. As we saw in the Map Drift (MD) algorithm, the subaperturing can occur only in one dimension, if needed.

8.2.2 Multilook for Speckle Reduction

Recall that speckle follows a negative exponential distribution for a SAR intensity image, where intensity is the sum of the squares, and not the square root of the sum of the squares, which is magnitude. This probability density function is defined as

$$p_I(I) = \begin{cases} \frac{1}{\bar{I}} e^{\frac{-I}{\bar{I}}}, & I \geq 0 \\ 0, & \text{otherwise} \end{cases} \tag{8.7}$$

where $\bar{I}$ is the mean intensity, $I = A^2$ is the power or intensity of the complex image A. Fluctuations of the intensity about the mean can be large, because the standard deviation of the speckle intensity σ_I is equal to the mean intensity. Because of this large fluctuation, speckle can create a visually distracting effect that can mask weak scatterers, interfere with sidelobes of brighter targets, and diminish the visual interpretability of the overall image.

There are several ways to filter speckle, reducing its granular effect and ultimately improving image interpretability. Multilook/subaperturing can be used to accomplish this. Multilook takes several subapertures of an image and incoherently adds their magnitude images. This incoherent addition averages out the noise, which will have a different random distribution in each subaperture, while increasing the SNR of the strong scatterers in the scene. The advantage is a visually pleasing image that is suitable for interpretation or exploitation. The disadvantage is the reduction of resolution in the scene, depending on the number of subapertures used. The subapertures can either be

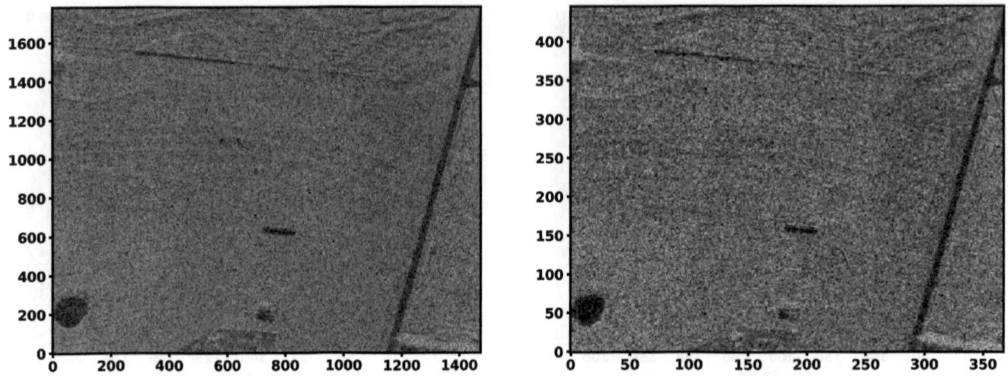

Figure 8.4 An MSTAR image of a rural area with a road and a small body of water in the lower left corner of the image.

overlapping or nonoverlapping. A common practice, if the SNR of the overall SAR system is such that speckle is a significant issue, is to collect at an azimuth and range resolution that is significantly finer than needed in the final image product, then to subaperture, detect to create the magnitude image, and then incoherently add. Once added, the image is contrast scaled in such a way to create a visually pleasing image.

Example 8.2 The image shown in Figure 8.4 was taken as part of DARPA's MSTAR program with a high-resolution X-band SAR system. The native resolution is 0.3 meter in both range and azimuth. The scene shows a road on the right-hand side of the scene with a large grassy field and a pond or other small body of water on the lower left side of the image. The image on the left is with no multilook applied, while the image on the right is the same image with a 16-look multilook. Sixteen subapertures were taken, nonoverlapping, from the original phase history and their magnitudes are averaged together to create a cleaner image. Note that the dark road and the pond are both cleaner than the original image. Both images were created using a "three-sigma" contrast stretch, which consists of setting the low and high values of the image to $[\mu - 3\sigma, \mu + 3\sigma]$, respectively.

The subapertures were created by taking a 4×4 grid of subapertures, not overlapping, from the original phase history, this implies that the resolution is now one-quarter of the original resolution in both azimuth and range, giving the new image a range and azimuth resolution of about 1.2 meters, instead of the 0.3-meter resolution that the sensor natively captured.

8.3 Other Speckle Reduction Techniques

Multilook is not the only speckle reduction technique. Argenti, Lapini, and Alparone (2013)[4] review several methods, including Bayesian estimators, probability density

[4] F. Argenti, A. Lapini, and L. Alparone, "A Tutorial on Speckle Reduction in Synthetic Aperture Radar Images," *IEEE Geoscience and Remote Sensing Magazine*, doi: 10.1109/MGRS.2013.2277512, September 2013.

function models, wavelet transforms, spatial filtering, and nonlocal filtering. One common filter is the Refined Lee filter.[5] The filter is simple in design, simple to implement, and fast to execute. It has the advantage of not significantly reducing resolution while cleaning up the image.

The Refined Lee filter relies on local statistics around a given pixel to build the filter. Let $z_{i,j}$ be the value (brightness) of a given pixel (i, j) in a 2D image. The variable $x_{i,j}$ represents the pixel before speckle is introduced (i.e., the idealized pixel). The mathematical model for additive noise is

$$z_{i,j} = x_{i,j} + w_{i,j} \tag{8.8}$$

where $w_{i,j}$ is a zero-mean Gaussian random sequence (mean of zero, standard deviation of σ). Lee estimated that a priori mean and variance of $x_{i,j}$ by assuming that you could use the local mean and variance of all of the pixels is a local neighborhood in the proximity of (i, j). Let

$$\bar{x}_{i,j} = \bar{z}_{i,j} \tag{8.9}$$

$$Q_{i,j} = E\left[\left(z_{i,j} - \bar{z}_{i,j}\right)^2\right] - \sigma_1^2 \tag{8.10}$$

where $\bar{x}$ represents the mean of x (or other variable under the bar), and $\bar{z}_{i,j}$ and $E\left[\left(z_{i,j} - \bar{z}_{i,j}\right)^2\right]$ are approximated by a local mean and variance of some window around (i, j). The variance σ_1^2 is estimated across the entire image. Lee showed[6] that the minimum mean square filter is

$$\hat{x}_{i,j} = \bar{x}_{i,j} + k_{i,j}\left(z_{i,j} - \bar{x}_{i,j}\right) \tag{8.11}$$

where $k_{i,j}$ represents a set of weights between $(0, 1)$ and defined as

$$k_{i,j} = \frac{Q_{i,j}}{Q_{i,j} + \sigma_1^2} \tag{8.12}$$

Thus, the Refined Lee filter uses a sliding window of size $N \times N$, where N is odd and (i, j) is at the center of the window. The local statistics are calculated, $\bar{x}_{i,j}$ (mean) and $Q_{i,j}$ (variance), and then the variance over the whole image σ_1^2 is calculated. Then the center pixel is replaced by the value of Equation (8.11). This is done for each pixel in the scene, where edge pixels are either ignored or a circular shift is used to link top edge pixels with bottom edge pixels and left edge pixels with right edge pixels.

In Lee (1981),[7] it is asserted that a 7×7 window is typically used. However, this is dependent on the original image size and resolution cell size. The neighborhood size can be increased or decreased to improve the local statistics; however, there is no specific rule-of-thumb to do this.

---

[5] J. S. Lee, "Refined Filtering of Image Noise Using Local Statistics," *Computer Graphics and Image Processing*, Vol. 15, No. 4, pp. 380–389, 1981.

[6] J. S. Lee, "Digital Image Enhancement and Noise Filtering by Use of Local Statistics," *IEEE Transactions on PAMI*, Vol. PAMI-2, No. 2, 1980.

[7] J. S. Lee, "Refined Filtering of Image Noise Using Local Statistics," *Computer Graphics and Image Processing*, Vol. 15, No. 4, pp. 380–389, 1981.

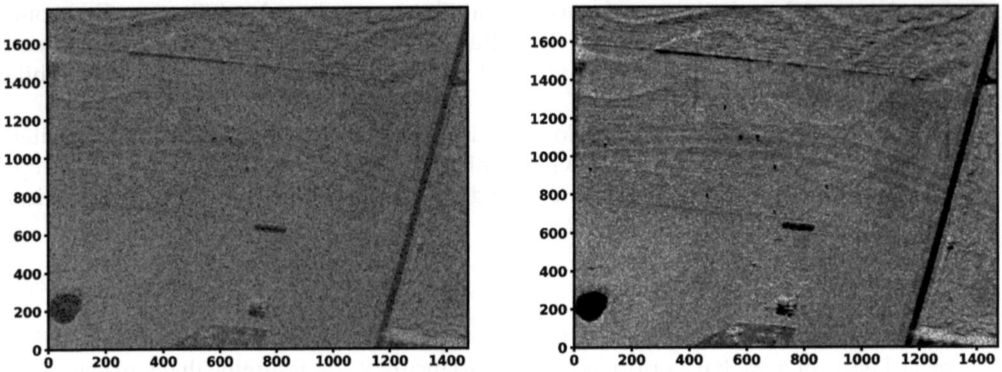

Figure 8.5 Speckle reduction using a 5 × 5 Refined Lee filter.

Example 8.3 The same MSTAR image is compared in Figure 8.5, with no speckle filtering on the left and Lee filtering on the right. The Lee filter uses a 5 × 5 sliding window to create the mean and variance images. Both images were displayed with a three-sigma stretch. The Lee filtered image has a much better contrast than the original image and, arguably, is of higher quality than the multilook image in the previous example. More importantly, the resolution of the Lee filtered image was not sacrificed to reduce the speckle noise. In particular, the texture of the scene appears to be better defined in the Lee filtered image.

8.4 Aperture Weighting

In Fourier analysis, a sinusoidal signal will leak if it is not perfectly centered on the digital Fourier transform (DFT) basis frequency. The leaking will push power from the main-lobe in the impulse response (IPR) into adjacent frequency bins. This effect of this leaking is referred to as *sidelobes*. Figure 8.6 shows an example IPR function with no windowing

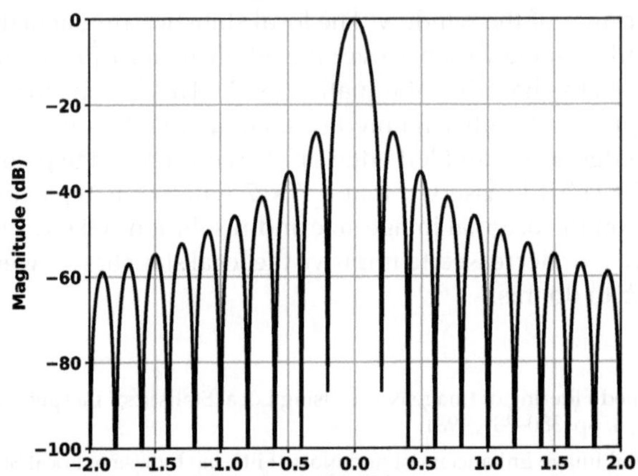

Figure 8.6 An example impulse response (IPR) function.

for a finite-length sequence of data (or, if you like, rectangular windowing, which is what you get when the windowing function is essentially unity for each sample). The mainlobe appears as the tallest peak in the center of the function. The peaks to either side of the mainlobe are the sidelobes and contain the remaining spectral energy. In some cases, particularly for a very strong scatterer, the sidelobes can interfere with mainlobes of weaker scatterers. It's desirable to apply some sidelobe control to emphasize the mainlobes and minimize the sidelobes. This is done through *windowing*. A window is a function that is applied in the time domain across all of the samples before a Fourier transform is applied to the data. Typically, for SAR, aperture weighting comes after phase adjustments are made, but often before spatial transformations, such as slant-to-ground plane transforms. Aperture weighting often sharpens an image visually.

There are many different windowing functions. Some of the common functions include Taylor, Hann, and Hamming windows. Each controls sidelobes in some way. Lower sidelobes mean a wider mainlobe, as a general rule. A wider mainlobe sacrifices some of the resolution in either range or azimuth, depending on the windowing function selected. There is no specific set of rules in selecting a windowing function; Taylor windowing is a common selection, as Taylor windowing approaches the ideal Dolph–Chebyshev characteristics of a minimally small mainlobe and uniform sidelobe levels.[8] The Dolph–Chebyshev window function has discontinuities at the edges, which can be undesirable, creating other spectral leakage effects. Taylor windowing is a suitable alternative, as it allows its equiripple to slowly decay as you get closer to the edge.

A rectangular window is defined mathematically as

$$w[n] = \begin{cases} 1, & 0 \leq n \leq M \\ 0, & \text{otherwise} \end{cases} \tag{8.13}$$

where M is the length of the sequence that needs windowing. A Hann window, developed by Julius von Hann,[9] is defined as

$$w[n] = \begin{cases} 0.5 - 0.5\cos\left(2\pi n/M\right), & 0 \leq n \leq M \\ 0 & \text{otherwise} \end{cases} \tag{8.14}$$

The Hamming window, named after Richard W. Hamming, is

$$w[n] = \begin{cases} 0.54 - 0.46\cos\left(2\pi n/M\right), & 0 \leq n \leq M \\ 0 & \text{otherwise} \end{cases} \tag{8.15}$$

The Taylor window is a bit more complicated, defined as

$$\bar{w}[\epsilon] = 1 + \sum_{m=1}^{\bar{n}-1} F_m \cos\left(2\pi m\epsilon\right), |\epsilon| \leq 1/2 \tag{8.16}$$

[8] C. V. Jakowatz, D. E. Wahl, P. H. Eichel, D. C. Ghiglia, and P. A. Thompson, *Spotlight-Mode Synthetic Aperture Radar: A Signal Processing Approach*, Kluwer Academic Publishers, Boston, MA, 1996.

[9] A. V. Oppenheim and R. W. Schafer, *Discrete-Time Signal Processing*, Prentice Hall Signal Processing Series, Englewood Cliffs, NJ, 1989.

The sequence $w[n]$ is obtained by taking N equally spaced points between $\epsilon = [-1/2, +1/2]$, so that $w[n] = \bar{w}[\epsilon[n]]$, where

$$\epsilon[n] = \frac{n + 1/2}{N} - \frac{1}{2}, n = 0, 1, \ldots, N-1 \tag{8.17}$$

The weights on the cosine function F_m is defined as

$$F_m = \frac{(-1)^{m+1} \prod_{n=1}^{\bar{n}-1} \left[1 - \frac{m^2/\sigma_p^2}{A^2 + (n-1/2)^2}\right]}{\prod_{n=1, n\neq m}^{\bar{n}-1} \left[1 - \frac{m^2}{n^2}\right]} \tag{8.18}$$

where A is influenced by the user defined sidelobe parameter S_L

$$A = \frac{1}{\pi} \cosh^{-1}\left(10^{S_L/20}\right) \tag{8.19}$$

The sidelobe parameter S_L expressed in decibels, determines where the first sidelobes should be in reference to the mainlobe. That is, if the mainlobe is at 0 dB and $S_L = 35$ dB, then the first sidelobes will end up at −35 dB. This can be seen in Figure 8.8, where a 35-dB Taylor window is used. The first sidelobes are at −35 dB from the peak of the mainlobe. The parameter σ_p is the ratio of the 3-dB down mainlobe width to the width of the ideal Dolph–Chebyshev half amplitude (at 3 dB).[10] This ratio is directly related to the mainlobe broadening factor and is dependent on A and the number of terms in the windowing function $\bar{n}$. This is defined as

$$\sigma_p = \frac{\bar{n}}{\sqrt{A^2 + (\bar{n} - 1/2)^2}} > 1 \tag{8.20}$$

The maximum amount of broadening of the mainlobe is defined as

$$\sigma_{p\max} = \frac{\sqrt{4A^2 + 1}}{2A} \tag{8.21}$$

when

$$\bar{n} \approx 2A^2 + 1/2 \tag{8.22}$$

As $\bar{n}$ gets arbitrarily large, we have the limit of σ_p approaching 1. Thus, the Taylor window approaches the ideal Dolph–Chebyshev's desirable characteristics as $\bar{n}$ approaches infinity. These characteristics include minimum mainlobe broadening and uniform, well-behaved sidelobe levels. Since an infinite number of terms in F_m are not possible, we have to bound $\bar{n}$. As long as $\bar{n} > 2A^2 + 1/2$, it represents a good approximation to the Dolph–Chebyshev. The rectangular, Hann, Hamming, and Taylor windows are shown in Figure 8.7, while the results of applying these windows to an ideal point response are shown in Figure 8.8.

---

[10] C. V. Jakowatz, D. E. Wahl, P. H. Eichel, D. C. Ghiglia, and P. A. Thompson, *Spotlight-Mode Synthetic Aperture Radar: A Signal Processing Approach*, Kluwer Academic Publishers, Boston, MA, pp. 170, 1996.

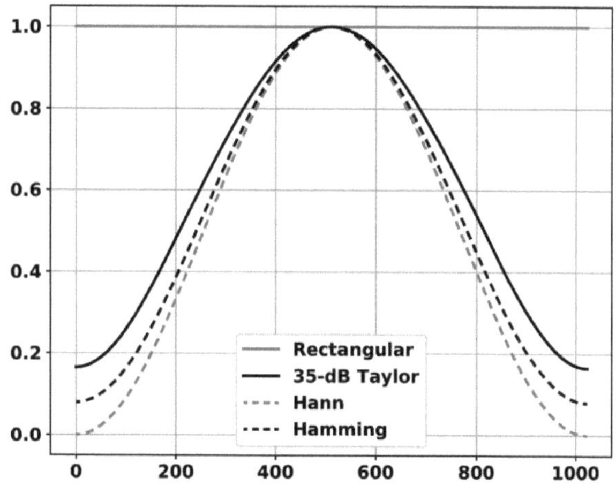

FIGURE 8.7 The windowing functions for rectangular, Hann, Hamming, and Taylor.

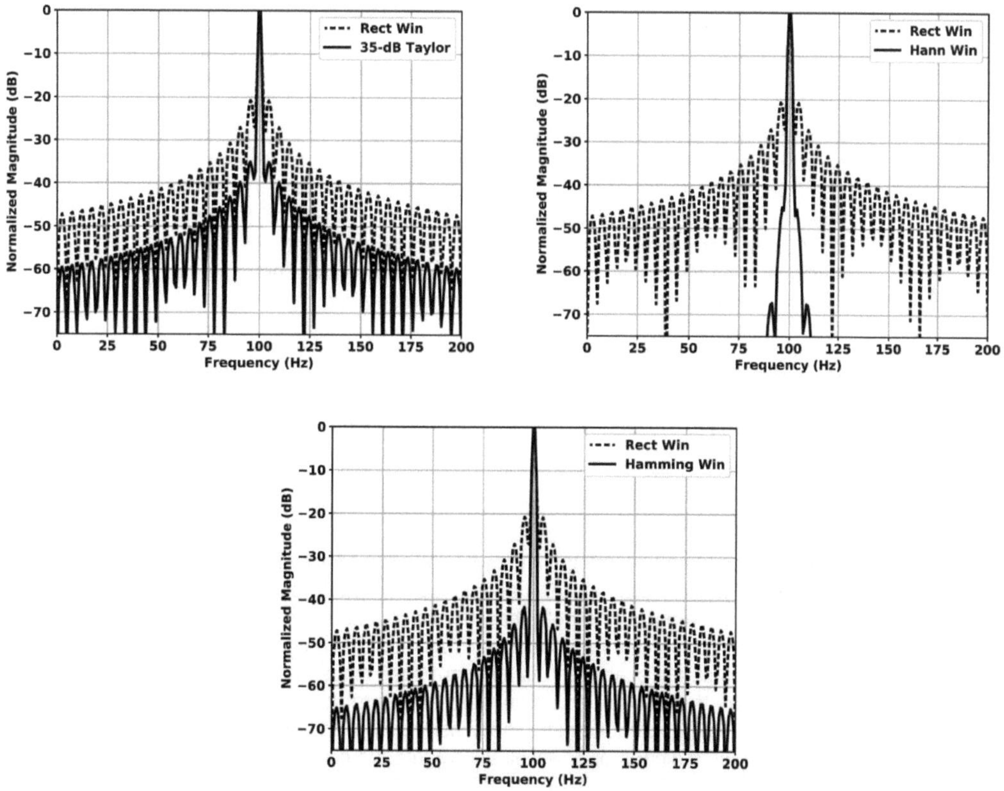

FIGURE 8.8 Results of windowing as compared to an IPR with no windowing/rectangular windowing.

Metric	Description	Typical Value
Impulse response width (IRW)	This width is the width of the mainlobe of the IPR, measured at the 3-dB down point. This should be equivalent to the image resolution in either the range or azimuth direction, depending on which slice through the IPR was measured. An illustration of the IRW can be seen in Figure 8.9.	1.0 m
Peak sidelobe ratio (PSLR)	This ratio is the height between the largest sidelobe (typically the first sidelobe) and the mainlobe (in decibels). Figure 8.9 illustrates the PSLR and how it is measured. For the example of the 35-dB Taylor window in Figure 8.8, the PSLR is −35 dB.	−18 dB
1D integrated sidelobe ratio (ISLR)	This metric measures the sidelobe power in either the range or azimuth direction. The integrated sidelobe ratio can be computed by integrating the power (intensity or magnitude squared) of the IPR over a region of the span of the IPR. This region is somewhat arbitrary, as the sidelobe power decays rapidly. If the power of the mainlobe is defined as P_{main} and the integrated power is the total power P_{total}, then the ISLR is $$\text{ISLR} = 10\log_{10}\left(\frac{P_{\text{total}} - P_{\text{main}}}{P_{\text{main}}}\right)$$	−17 dB
2D integrated sidelobe ratio (ISLR)	The 2D ISLR is defined similarly to the 1D ISLR, with the entire scatterer defined in a 2D rectangle 1.2–1.4 times the respective resolutions. The sidelobe power is thus defined as the power of the entire 2D chip less the mainlobe power. However, a simpler approximation is to measure the 1D ISLR in both directions. This assumes the 2D impulse response to be separable in both directions $p(t_1, t_2) = p_1(t_1, B_1)p_2(t_2, B_2)$, where $p_i(t_i, B_i)$ is a sinc-squared-like function with bandwidth B. If there is no weighting, then $p_i(t, B) = \text{sinc}(\pi Bt)$ for a baseband signal. Cumming and Wong state that if the ISLR is equal in both directions then $\text{ISLR}_{2d} \approx \text{ISLR}_{1d} + 3$.	−14 dB
Peak position	This is the position of the peak of the IPR in image space. This metric is used for geometric calibration; this requires understanding whether the image space is in the slant plane or the ground plane (described in the next section).	2 m
Signal peak magnitude	Used for radiometric calibration, this metric quantifies the magnitude of the target response at its peak.	Target dependent
Phase	As we will see in an upcoming chapter, it's important for phase to preserved. This is critical for applications such as interferometry and polarimetry. It's important to note that the phase characteristics of a scatterer can varying in either direction. However, it is important to note that the compressed scatterer has a linear phase across the peak of the IPR, where the slope is proportional to the center frequency of the data in that direction. Additionally, the consistency of the phase in both directions is an indicator of how well the targets are compressed.	0.3 radians

Table 8.1 Quantitative Image Quality Measures

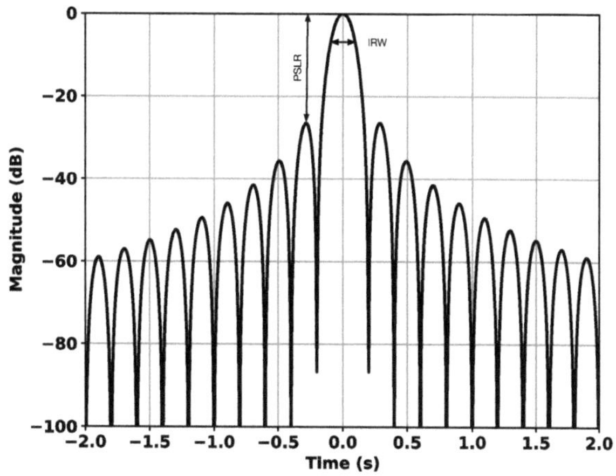

Figure 8.9 Two of the image quality metrics, impulse response width (IRW) and peak sidelobe ratio (PSLR), illustrated on an ideal impulse response of a scatterer.

8.5 Image Quality Metrics

While it is common for a person to look at an image and qualitatively judge its quality ("its sharp" or "its blurry" or "its contrast is too high"), this is not particularly useful for comparing images or analyzing changes to the image processing chain. Quantitative measures are preferred for comparing images to one another or for doing regression testing on a single image when making changes to the processing software. While the IPR of a given scatterer is two-dimensional, it is often useful to examine the IPR as a slice in either azimuth or range (or both, separately). Cumming and Wong (2005)[11] have an excellent set of quantitative metrics; they are summarized in Table 8.1.

[11] I. G. Cumming and F. H. Wong, *Digital Processing of Synthetic Aperture Radar Data: Algorithms and Implementation,* Artech House, Norwood, MA, pp. 60–63, 2005.

CHAPTER 9

Linear Frequency-Modulated Continuous Wave Systems

In recent years, there has been a good amount of research and development in the field of very small, low-cost, low-power synthetic aperture radar (SAR) sensors. Many of these systems employ a linear frequency-modulated continuous wave (LFMCW) waveform, rather than the pulsed waveform discussed in the earlier chapters of the book. This chapter will introduce the theory behind LFMCW systems, as well as some of the difficulties associated with processing the data to high-resolution, high-quality images.

9.1 LFMCW Systems

Up until now, we have been discussing pulsed radar systems for synthetic aperture radar. In these pulsed systems, the radar is treated as a "stop and hop" system, where each pulse is transmitted and received as if the sensor is stationary, the sensor moves to the next position, and then the next pulse is transmitted and received. This model holds true for high-altitude aircraft and spacecraft. However, as the sensor gets close enough to the ground, the round-trip delay time gets shorter and shorter, making a pulsed system more and more difficult to design and implement. Instead, for short-range synthetic aperture radars, a continuous train of pulses can be transmitted, received, demodulated, and processed to create the SAR image. The hardware to do this is not particularly complicated and, in some cases, can be inexpensive. However, there are new issues that have to be considered, particularly, how motion compensation is performed, particularly since there is a true Doppler effect induced on the data since it is a continuous wave system.

Figure 9.1 shows two different types of LFMCW models. The first, shown in the top plot, is a triangle wave. Note that both plots are in frequency versus time. The sensor is constantly transmitting an up chirp then a down chirp. The second model, a sawtooth wave, the sensor continuously sends up chirps. This creates a discontinuity between

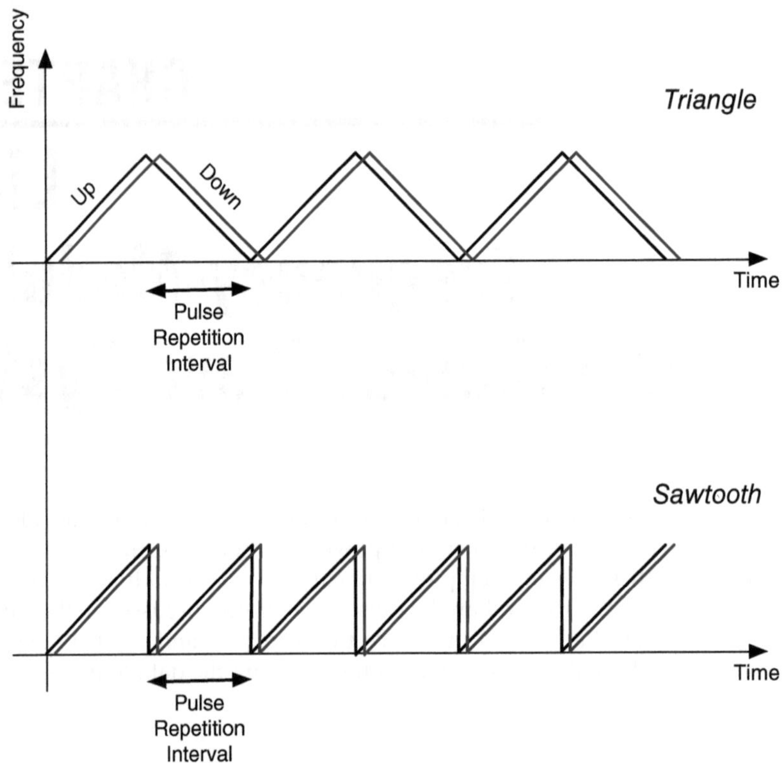

FIGURE 9.1 Two examples of linear frequency-modulated continuous wave (LFMCW) waveforms: a triangle wave and a sawtooth wave.

chirps that can generate higher frequency noise, which may have to be filtered either in the receiver or in the post-processing stage. The gray lines represent a time delayed return from some arbitrary scatterer in the scene.

Each up or down chirp is treated, in the transmitter-receiver and in image formation processing, as a discrete pulse. This makes handling the data more manageable. The period of time between the start and end of an (up/down) chirp is equivalent to the *pulse repetition interval* (PRI). We use this terminology less for accuracy (i.e., there are no pulses) and more for consistency. As before, the inverse of the PRI is the *pulse repetition frequency* (PRF). Again, as before, the selection of the PRF in system design is part science and part art form.

9.2 LFMCW Transmitter-Receiver Model

Figure 9.2 shows the basic system model for an LFMCW system. The transmitter is represented by the top line of boxes, whose flow is from left-to-right. The receiver is shown by the bottom line of boxes, whose flow is from right-to-left. The system has two antennas, one for transmit and the other for receive.

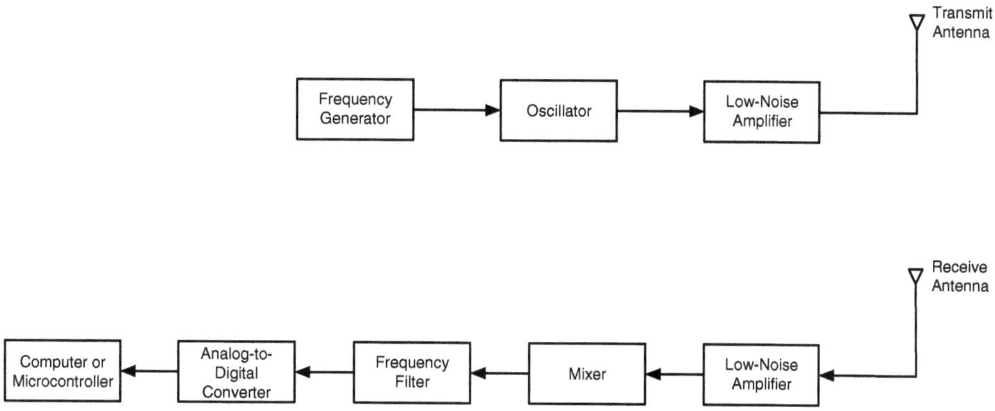

Figure 9.2 The basic system model for a simple LFMCW radar.

Example 9.1 Consider an LFMCW system mounted on a unmanned aerial vehicle, such as a quadcopter. If the quadcopter flies at $h = 50$ m above ground level (AGL) and the grazing angle of the antenna is fixed at $\psi = 60°$, then the range between the antenna reference point (ARP) and the ground reference point (GRP) is

$$R = \frac{h}{\sin(\psi)} = \frac{50 \text{ m}}{\sin(60°)} = 57.7 \text{ m} \tag{9.1}$$

At a range of 57.7 m, the round-trip delay time is

$$\tau = \frac{2R}{c} = \frac{2 \cdot 57.7 \text{ m}}{299792458.0 \text{ m/s}} = 385 \text{ ns} \tag{9.2}$$

The transmitter generates a train of LFM pulses, typically either in triangle or saw-tooth form, using the function generator and an oscillator [typically a voltage-controlled oscillator (VCO)]. The VCO then creates the linear frequency-modulated chirps at the radar center frequency. The train of pulses is sent out. The receiver is typically turned on and left on during the image stage, unlike in pulsed radar where only samples during the listening window correspond to the time when returns are expected from the ground swath. The receiver captures the return signals, amplifies that signal, mixes the signal with the reference signal (the signal directly from the transmitter), and then filters the signal for the intermediate frequencies of interest. That signal is then digitized and stored for image formation processing.

The systems that perform this range from the simplistic[1] to the elaborate,[2] particularly for do-it-yourself radar. The next section considers the mathematical model for

[1] G. L. Charvat, J. H. Williams, A. J. Fenn, S. Kogon, and J. S. Herd, "Build A Small Radar System Capable of Sensing Range, Doppler, and Synthetic Aperture Radar Imaging," https://ocw.mit.edu/resources/res-11-003-build-a-small-radar-system-capable-of-sensing-range-doppler-and-synthetic-aperture-radar-imaging-january-iap-2011/

[2] http://hforsten.com/synthetic-aperture-radar-imaging.html

LFMCW systems and how the stop-and-hop assumption for pulsed radar systems no longer holds.

9.3 LFMCW Mathematical Model

There has been much published recently in the popular science press and in peer-reviewed journals about linear frequency-modulated continuous wave systems. However, the work published by Evan Zaugg and David Long[3,4,5,6,7,8,9] from Brigham Young University stands out. The mathematical model used in Zaugg and Long (2015) will be summarized here.

Recall that Euler's formula states that

$$e^{jx} = \cos(x) + j\sin(x) \tag{9.3}$$

This exponential notation is easier to write and comprehend, as it encapsulates both the I and Q channels that we've been discussing. For these low-cost, low-power LFMCW systems, there is typically no Q channel. Only the I channel is dechirped and then digitized. The Q channel can be created using a Hilbert transform.[10]

Consider a single transmitted linear frequency-modulated "up" chirp

$$s_t\left(\eta, t\right) = e^{j\left(\phi + 2\pi f_0 t + \pi K t^2\right)} \tag{9.4}$$

where ϕ is the initial phase at the start of the chirp, f_0 is the center frequency of the radar, K is the chirp rate, η is the independent slow time variable, and t is the independent fast

[3] E. C. Zaugg, "Generalized Image Formation for Pulsed and LFM-CW Synthetic Aperture Radar," Theses and Dissertations, 2010.

[4] E. Zaugg, D. Long, M. Edwards, M. Fladeland, R. Kolyer, I. Crocker, J. Maslanik, U. Herzfeld, and B. Wallin, "Using the MicroASAR on the NASA SIERRA UAS in the Characterization of Arctic Sea Ice Experiment," *Radar Conference*, pp. 271–276, 2010.

[5] E. Zaugg, M. Edwards, and A. Margulis, "Developing a Small Multi Frequency Synthetic Aperture Radar for UAS Operation: the SlimSAR," *presented at the SPIE Defense, Security, and Sensing, 2010*, Vol. 7669, pp. 766902-1 – 766902-9.

[6] M. C. Edwards, E. C. Zaugg, D. G. Long, R. Christiansen, and A. Margulis, "A Small, Manned Aircraft as a Testbed for Radar Sensor Development," *presented at the SPIE Defense, Security, and Sensing, 2009*, Vol. 7308, pp. 730807–730807–7.

[7] E. C. Zaugg, D. L. Hudson, and D. Long, "The BYU SAR: A Small, Student-Built SAR for UAV Operation," *presented at the IEEE International Conference on Geoscience and Remote Sensing Symposium, 006, IGARSS 2006*. 2006, pp. 411–414.

[8] E. C. Zaugg and D. G. Long, "Theory and Application of Motion Compensation for LFM-CW SAR," *Geoscience and Remote Sensing*, Vol. 46, No. 10, pp. 2990–2998, 2008, doi: 10.1109/36.298008.

[9] E. C. Zaugg and D. G. Long, "Generalized Frequency Scaling and Backprojection for LFM-CW SAR Processing," *IEEE Transactions on Geoscience and Remote Sensing*, Vol. 53, No. 7, pp. 3600–3614, March 2015, doi: 10.1109/36.298008.

[10] A-H. Najmi and T. K. Moon, "Advanced Signal Processing: A Concise Guide," McGraw Hill, New York, 2020.

time variable. A "down" chirp is defined as

$$s_t\left(\eta, t\right) = e^{j\left(\phi+2\pi(f_0+B)t-\pi Kt^2\right)} \tag{9.5}$$

where B is the bandwidth of the radar. For the sake of introduction to the mathematical model, we will only consider up chirps. If τ is the round-trip time delay between the radar and an arbitrary scatterer, then at some time t, the radar is receiving the signal reflected from that scatterer at $t-\tau$. The range to that scatterer is estimated using τ. However, the actual time delay between t and $t-\tau$ is different due to the constant motion of the aircraft. As has been discussed, pulses systems use a stop-and-hop assumption to approximate τ, and thus, range. Because of this, τ is a function of the slow-time variable η. However, in a continuously transmitting system, τ is actually a function of t as well as η. The constant motion induces a change in range over time. In many cases, this difference in range can be neglected. Because of this, the relationship between τ and range R is still

$$\tau \approx \frac{2R(\eta, t)}{c} \tag{9.6}$$

where $R(\eta, t)$ is defined as

$$R(\eta, t) = \sqrt{R_0^2 + v_a^2(t+\eta)^2} \tag{9.7}$$

where R_0 is the distance of closest approach to some arbitrary scatterer. This relationship now accounts for the change in range in both fast time and slow time. With this relationship in mind, the received signal from some scatterer at $R(\eta, t)$ is

$$s_r(\eta, t) = e^{j\left(\phi+2\pi f_0(t-\tau)+\pi K(t-\tau)^2\right)} \tag{9.8}$$

Note that the signal does not indicate that there is an attenuation in the signal as it is transmitted, reflected, and then received in the radar. This is only to simplify the mathematics; the signal from an arbitrary scatterer is still an attenuated, time-delayed chirp when it gets to the receive antenna.

The received signal, Equation (9.8), is mixed with the reference signal, which is the complex conjugate of Equation (9.4). The resulting equation takes the form

$$s_{dc0}(\eta, t) = e^{j\left(2\pi f_0\tau+2\pi Kt\tau-\pi K\tau^2\right)} \tag{9.9}$$

Zaugg and Long (2015) warn that this method, called *direct dechirp*, is limited because it requires sampling all of the distance from the start of the chirp all the way to the end of the chirp, which may correspond to a significant amount of empty space between the SAR and the ground. They describe a mathematical model that delays the reference signal by some time d. However, such an analog delay may be costly, both in cost, weight, and power. Such a delay may be relatively easy to implement in a software defined radio implementations of LFMCW.

Figure 9.3 illustrates how the dechirp process for a direct dechirp system works. In this figure, the continuous wave is a triangle wave with alternating up and down chirps. The scale for the illustration is exaggerated to make the distinction between the transmitted and received signals. The transmitted signal is shown in black, while the delayed signal is shown in gray. The return signal is delayed by τ seconds. The dechirp

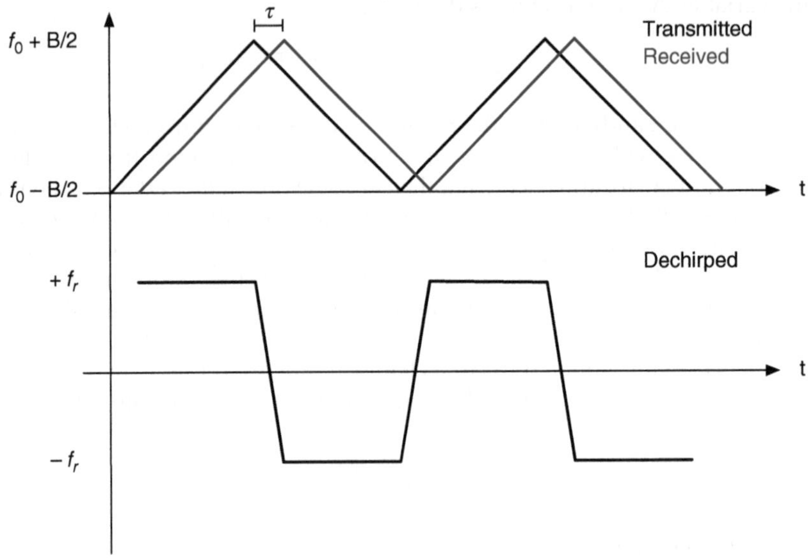

FIGURE 9.3 A graphical description of how dechirp works in LFMCW systems.

process creates a single sinusoid for each scatterer (ignoring, temporarily, the change in frequency induced by the Doppler effect due to the motion of the aircraft). Each frequency for each sinusoid in the superposition of sinusoids is directly related to range by

$$f_r = \frac{2RK}{c} \tag{9.10}$$

which is directly derived from the second term in Equation (9.9). The first term in that equation is just a phase shift based on the position of the scatterer in the scene with relation to the antenna reference point. The third term is akin to the range skew discussed in Chapter 6.

The sampling frequency f_s is directly related to the maximum range $R_{\max}$ and Equation (9.10),

$$f_s > \frac{2R_{\max}K}{c} \tag{9.11}$$

Again, recognize that this sampling frequency assumes that the signal is bandlimited and complex.

There is a transition period between chirps where the frequency for a given scatterer goes from $+f_r$ to $-f_r$ over τ seconds. In practice, depending on the range from the aircraft to the scatterer, this linear frequency-modulated signal will not create any issues, as τ will be much smaller than the sampling interval. However, if this is a concern, a range deskew can be applied to account for the transitory period between scatterers at the front edge and scatterers at the far edge of the swath.

9.4 Motion Compensation for LFMCW

To understand how motion of the platform affects the continuous wave signal, Zaugg and Long (2015)[11] take the raw phase history data and, for each pulse, take the Fourier transform so that the data is "range compressed." In doing so, they point out that the data in the range-compressed space (what we've also called the range-Doppler space) takes on the form

$$S_R = \int_{\frac{2R}{c}}^{\frac{2R}{c}+T_p} s_{dc}(\eta, t) e^{-j2\pi f_r t} dt \tag{9.12}$$

$$\approx T_p \operatorname{sinc}\left[T_p\left(f_r + Kd - K\tau\right)\right] e^{j\pi\Phi_{rc}} \tag{9.13}$$

where

$$\Phi_{rc} = \frac{2KRT_p}{c} - f_r T_p + \frac{4KR^2}{c^2} + \frac{4f_0 R}{c} - \frac{4f_r R}{c} \tag{9.14}$$

where R is the range to a given scatterer and it is assumed that R is not a function of t, fast time f_r is the frequency associated with that range, K is the chirp rate, and T_p is the pulse duration.

Assuming a simple linear flight path with constant velocity, the change in range can be approximated as

$$\tau(\eta, t) = \frac{2\sqrt{R_0^2 + v^2(\eta + t)^2}}{c} \tag{9.15}$$

$$\approx \frac{2R(\eta, 0)}{c} + \frac{2v^2 \eta t}{R(\eta, 0)c} \tag{9.16}$$

where

$$R(\eta, 0) = \sqrt{R_0^2 + v^2\eta^2} \tag{9.17}$$

and R_0 is the range of closest approach.

By taking this approximation for τ and putting into the term for $s_{dc}(\eta, t)$, it's clear that

$$s_{dcm}(\eta, t) = s_{dc}(\eta, t) e^{j\Phi_m} \tag{9.18}$$

where

$$\Phi_m = \left(-\frac{4v^4\eta^2\pi K}{R^2c^2} + \frac{4v^2\eta\pi K}{Rc}\right)t^2 + \left(-\frac{8v^2\eta\pi K}{c} + \frac{4v^2\eta\pi f_0}{Rc}\right)t \tag{9.19}$$

That is, there is an extra phase term added to s_{dc} in the form of Φ_m. Zaugg and Long point out that this extra phase term has an effect on a given scatterer: it shifts the scatterer

[11] E. C. Zaugg and D. G. Long, "Generalized Frequency Scaling and Backprojection for LFM-CW SAR Processing," *IEEE Transactions on Geoscience Remote Sensing*, Vol. 53, No. 7, July 2015.

in range, broadens its mainlobe, and raises the sidelobes. They demonstrate that the Doppler frequency over slow time is

$$2\pi f_d = 2\pi 2v \frac{v\eta}{R} \frac{f_0}{c} \tag{9.20}$$

They show that Equation (9.19) can be rearranged in terms of f_d, and neglecting the small terms, the phase correction to remove the Doppler is

$$\Phi_{\text{cor}} = -2\pi f_d t \tag{9.21}$$

This correction essentially removes the shift in range and restores the IPR to its proper width and sidelobe levels. In the rest of the work, Zaugg and Long describe how the motion compensation term can be used with backprojection and frequency scaling image formation to restore image quality when motion is noted to be nonlinear with varying velocity over the dwell. For backprojection in particular, they note that the range-compressed signal takes on the form in Equation (9.13). They recommend applying the correction in Equation (9.21) to correct for the effects of the continuous motion. After that correction is applied, then the standard Backprojection Algorithm (BPA) can be applied to focus the LFMCW SAR image.

Example 9.2 Consider a hypothetical C-band low-power, low-weight, low-cost LFMCW system with the parameters shown in Table 9.1.

With these parameters, the range and azimuth resolutions are approximately 0.5 meter. The swath is about 50 meters wide in range, while the azimuth width is about 73 meters. The chirp rate k_r is 1.5 GHz/s and the pulse width is 2 ms. Using Equation (9.10), the starting beat frequency which represents returns from the near edge of the swath is 122 kHz, while returns from the far edge will have a beat frequency of 174 kHz. For a dwell time of a half second and pulse repetition frequency of about 500 Hz, the number of azimuth samples equals 250, while the range samples along the

Parameter	Variable	Value
Center frequency	f_0	5.4 GHz
Transmitted bandwidth	R	300 MHz
Antenna beamwidth	β	20°
Grazing angle (at center of swath)	ψ	45°
Sensor altitude	z_{ac}	100 m
Sensor velocity	v_{ac}	15 m
Squint angle	θ	0°
Pulse repetition frequency	PRF	498.8 Hz
A/D sampling frequency	f_s	1 MHz
Dwell time	η_{tot}	0.5 s

Table 9.1 System Parameters for Hypothetical C-Band LFMCW Synthetic Aperture Radar

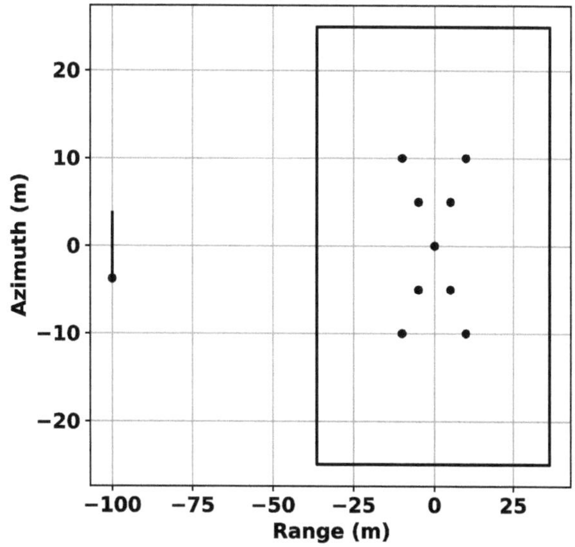

Figure 9.4 Collection geometry for hypothetical C-band LFMCW system.

pulse is equal to 2005. The range samples will be decreased to 1002 after the Hilbert transform is applied to make the real data collected from the system coherent.

Given these parameters, the collection geometry in the X–Y plane is shown in Figure 9.4. The flight line is shown on the far left of the graph, where the dot at the bottom of the line represents the sensor starting point and the line is the distance traveled over the dwell (0.5 second). The box represents the illuminated swath over the dwell and the dots inside the box represent idealized point targets. The nine point targets are in the configuration of an "X," centered at the GRP at $(0, 0, 0)$. Each of the targets lay on the ground at $z = 0$ meters.

Figure 9.5 shows the change in range and beat frequencies over slow time for each of the idealized point targets with respect to the sensor position. The ranges, shown in the graph on the left, are between 135 and 149 meters. This corresponds to range frequencies, also called beat frequencies, of 135 to 149 kHz.

Figure 9.6 shows the raw radar signal once the signal has been received, dechirped, and then converted to a digital signal. This is a real-valued signal and is not yet phase coherent. Figure 9.7 shows the first pulse received by the radar; this pulse has had dechirp applied and then been digitized. The values are shown as floating point values in the simulation, but in practice, the values are likely encoded into a signed integer. The superpositions of sinusoids and at their respective frequencies are directly related to the ranges of each of the nine targets at the very start of the collection. These sinusoids change frequencies slowly over slow time. It is that change in frequency that, as shown in previous chapters, can be used to create the synthetic aperture and generate the synthetic aperture radar image.

The Hilbert transform can now be applied to the raw radar data, creating coherent data. The data is now complex valued, representing the I and Q channels that were described in Chapter 2. Taking a 1D FFT along with each of the pulses will transform the data into the range-Doppler domain, where the data has been resolved to

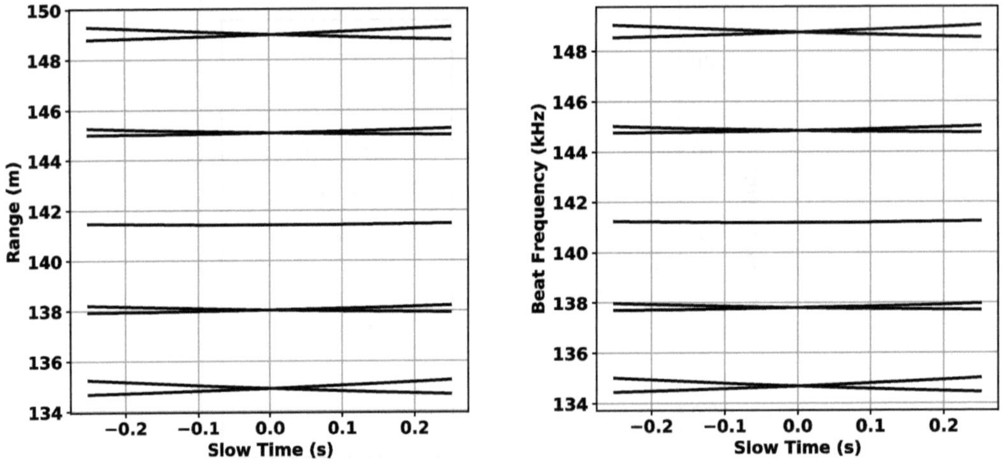

Figure 9.5 The ranges (left) and beat frequencies (right) for each of the idealized point targets with respect to the sensor position over slow time.

range but slow time (or Doppler, if you will) has been untouched. Examination of this data will reveal that the left half of the array (if the array has been properly center adjusted after the FFT) is set to zero, which is part of the Hilbert transform. This can be safely discarded, allowing for the remaining data to be scaled between $[0, f_s]$. The magnitude of the resulting complex radar data is shown in Figure 9.8. Examination of Figure 9.9 reveals that the real component of the first pulse of the complex radar

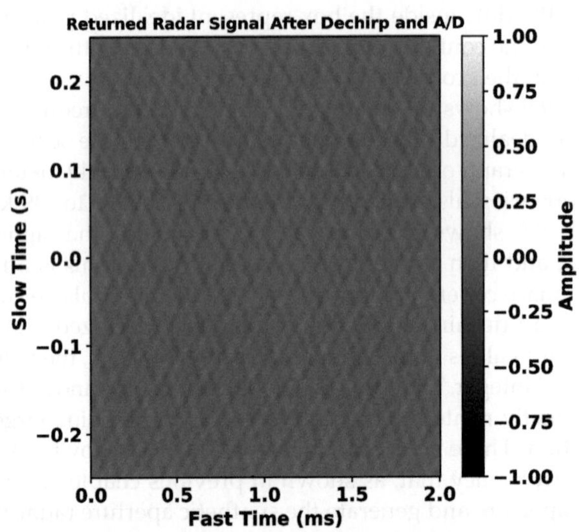

Figure 9.6 The raw radar return after dechirp and digitization. The pulses are stacked in the slow time direction (i.e., each pulse is a row in the image). At this stage in processing, the raw radar signal is still real valued and not yet phase coherent.

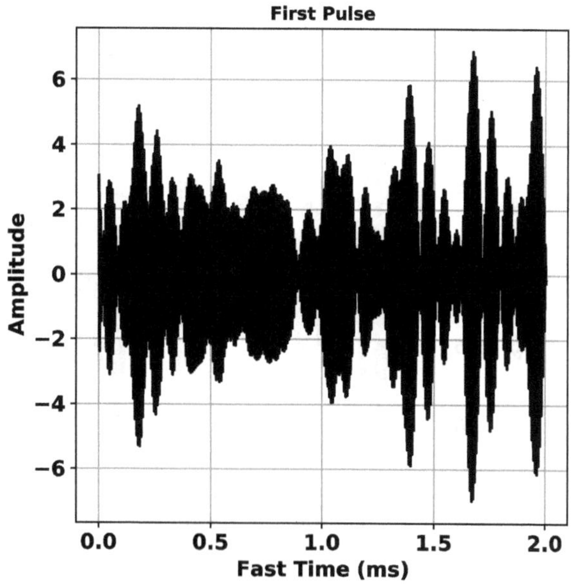

Figure 9.7 The first pulse captured by the radar, after dechirp and digitization.

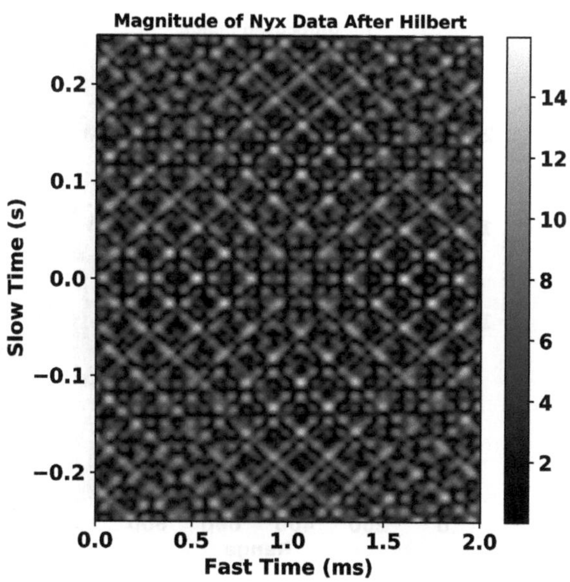

Figure 9.8 The magnitude of the complex radar data after the Hilbert transform is applied to the real-valued data.

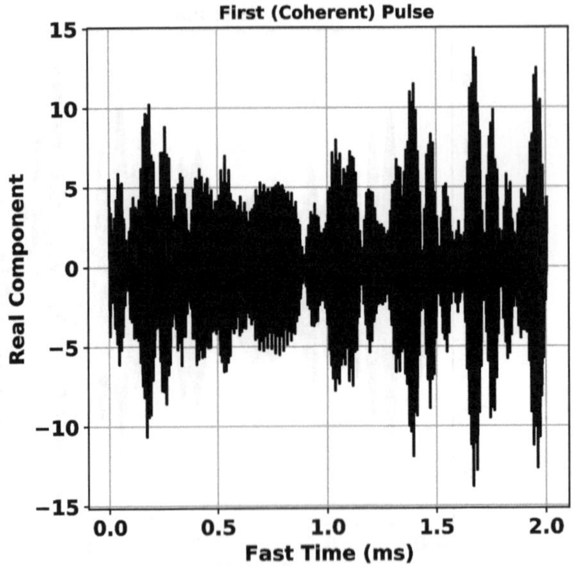

Figure 9.9 The real component of the first pulse of the complex radar data.

data maintains the same shape as the real-valued data before the Hilbert transform (compare to Figure 9.7).

A two-dimensional Fourier transform will take the raw radar data and put it in the "unfocused" image domain. The magnitude of the complex data is shown in Figure 9.10, displayed in decibels. The "X" shape of the targets is off-center in the

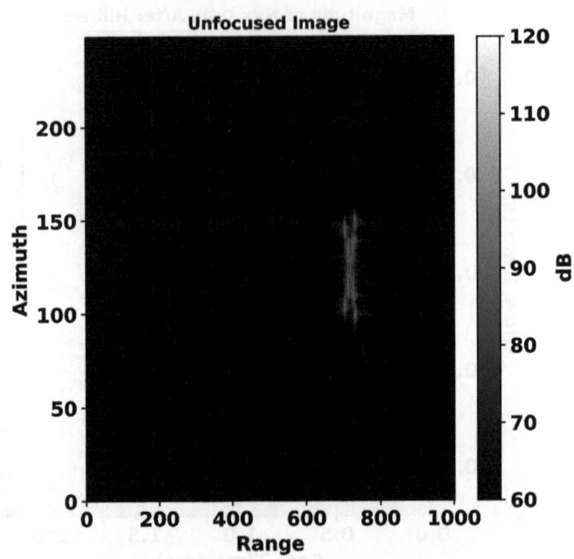

Figure 9.10 The magnitude image of the unfocused complex radar data, after a 2D FFT. Note the extreme distortion due to range migration.

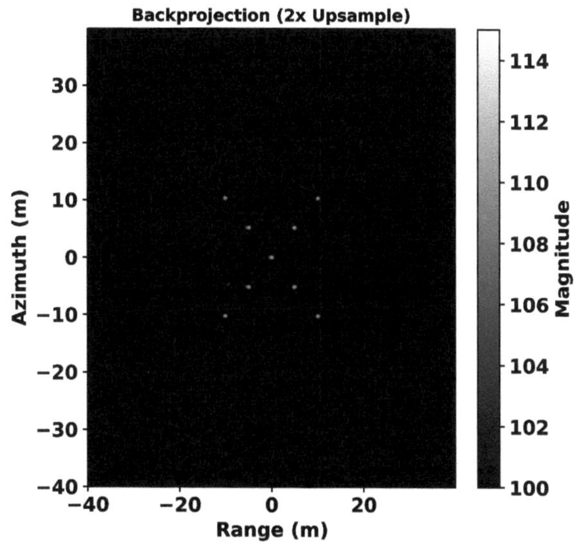

FIGURE 9.11 The magnitude image after Backprojection Algorithm (BPA) is used for image formation.

scene and highly distorted due to the range curvature. Because of the relative closeness of the sensor to the ground (as compared, say, to a satellite or high-altitude aircraft), the range curvature is significantly more pronounced. Simpler image formation techniques, such as Range-Doppler Algorithm (RDA) or Polar Formatting (PFA), may not completely correct the range migration. Omega-K, with its Stolt interpolation, could accurately correct the range migration, as would backprojection.

After the Doppler effect was corrected,[12] a 80 × 80 meter grid, centered around the GRP, was generated and BPA was applied. An upsampling factor of 2 and a Taylor windowing of 5 dB were used to generate the subsequent image. The image, shown in Figure 9.11, is now well-focused and centered at the GRP. Motion compensation is inherently part of the algorithm, as the ranges for each pixel in the 80 × 80 meter grid are calculated with respect of the distance to the sensor is updated for each processed pulse.

[12] E. C. Zaugg and D. G. Long, "Generalized Frequency Scaling and Backprojection for LFM-CW SAR Processing," *IEEE Transactions on Geoscience and Remote Sensing*, Vol. 53, No. 7, pp. 3600–3614, July 2015, doi: 10.1109/TGRS.2014.2380154

CHAPTER 10

Remote Sensing with Synthetic Aperture Radar

Up to this chapter, this work has focused on the creation of a well-formed, well-focused synthetic aperture radar (SAR) image. As has been discussed, these are called single-look complex (SLC) images and have been treated as microwave reflectivity pictures of a given scene. However, part of the power of SAR its is coherent nature. Because of this coherent nature, SAR can do much more than make an all-weather, any-time-of-day picture of a scene. Because of the coherent nature of SAR, as well as its unique collection geometry, there are a couple of unique products that can be generated out of the SLC data. The most common is arguably *interferometry* (IFSAR). Chapter 11 is dedicated to the description of IFSAR and its applications. Because of the assumption that the scene must be stationary and flat, objects that move, such as ground vehicles and ships, generate unique signatures because they violate those assumptions. Chapter 12 examines the phase errors induced by moving objects and how those phase errors can cause localized smearing and displacement of an object.

This chapter examines several other remote sensing applications of SAR, including polarimetry, ocean remote sensing, ice monitoring, oil spill detection, and automatic target recognition (ATR) using deep learning. Some of these applications are unique to SAR, having no analog in optical remote sensing.

10.1 Polarimetry

Up to now, we have only described electromagnetic (EM) waves in terms of their frequencies and phases. In truth, EM waves also contain directionality. This is called *polarization*. Polarization describes the direction a wave is traveling with respect to the antenna's z-axis. That is, if the wave is traveling up and down with respect to that z-axis, this is called *vertical polarization*. Conversely, if the wave is traveling side-to-side, it is called *horizontal polarization*. Figure 10.1 illustrates both vertical and horizontal polarization.

In the operation of a synthetic aperture radar, it is often the case that the transmit and receive polarization can be the same or different. If they are the same, this is called *co-pol* and is designated "HH" or "VV", for transmit and receive horizontally or vertically, respectively. If they are different, this is called *cross-pol* and is designed "HV" (for transmit horizontally and receive vertically) or "VH" (for transmit vertically and receive horizontally). If a system is capable of only transmitting one polarization, both receiving both

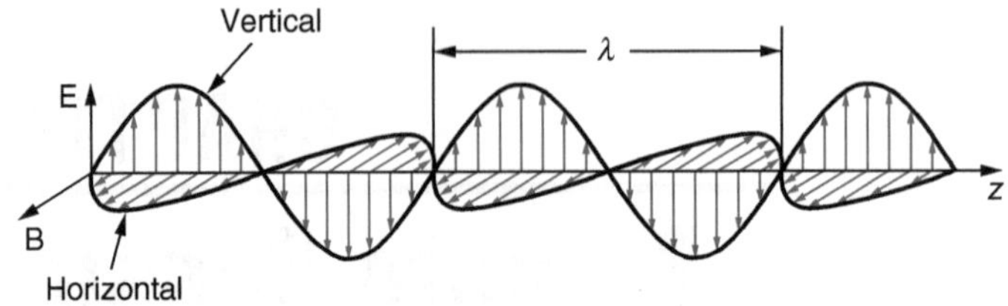

Figure 10.1 An illustration of both vertical and horizontal polarization. The vertical polarization is traveling up and down (like waves on the ocean) with respect to the *z*-axis, while the horizontal polarization travels orthogonal to the vertical.

polarization, this is called *dual pol*. If a system is capable of simultaneously collecting and receiving, HH, VV, HV, and VH, this is called *full pol* or *quad pol*.

Materials reflect polarized energy differently. Because of this, polarization can be used to separate out different types of materials, land use types, and ice. For example. in 1995, Hess et al.[1] demonstrated the delineation of floodplain inundation and vegetation in Brazil using polarimetric data from the SIR-C missions. These data, taken in April and October 1994, were used to classify the scenes into five different categories of land type: water, pasture, floating meadow, nonflooded forest, and flooded forest. Additionally, data from both the C-band and the L-band were leveraged to distinguish the different types, making their technique multifrequency and polarimetric.

Another application of polarimetry is ship detection. Liu, Vachon, and Geling[2] used features derived from the so-called scattering matrix to significantly improve ship detection in airborne SAR. Using quad pol data, they were able to show reduction in the probability of false alarms by two to three orders of magnitude.

Polarimetric images are often displayed as faux color images, where HH, VV, and HV (or VH) are used in the red, green, and blue color channels. This faux color image can then be used by an analyst to distinguish between certain types of materials and land uses. Another common application of polarimetry, and particularly of faux color imagery, is in agriculture. In a 2019 paper, Liu et al.[3] describe how synthetic aperture radar, and polarimetry in particular, can be used for the monitoring of crops. Polarimetric SAR can effectively measure the geometrical structures associated with different crops (e.g., corn versus soybean). Liu et al. describe three different applications: crop identification (types of crops being planted and/or harvested), crop parameter extraction (what size are the fields in which the crops are planted), and crop yield estimation (how much product will the field produce). They also identify some of the challenges in using SAR to monitor crops: spatial resolution requirements and SAR revisit rates (particularly for satellite SAR).

[1] L. L. Hess, J. M. Melack, S. Filoso, and Yong Wang, "Delineation of Inundated Area and Vegetation Along the Amazon Floodplain with the SIR-C Synthetic Aperture Radar," *IEEE Transactions on Geoscience and Remote Sensing*, Vol. 33, No. 4, pp. 896–904, July 1995.

[2] C. Liu, P. W. Vachon, and G. W. Geling, "Improved Ship Detection with Airborne Polarimetric SAR Data," *Canadian Journal of Remote Sensing*, Vol. 31, No. 1, p. 10, 2005.

[3] C. Liu, Z. Chen, Y. Shao, J. Chen, T. Hasi, and H. Pan, "Research Advances of SAR Remote Sensing for Agriculture Applications: A Review," *Journal of Integrative Agriculture*, Vol. 18, No. 3, pp. 506–525, March 2019.

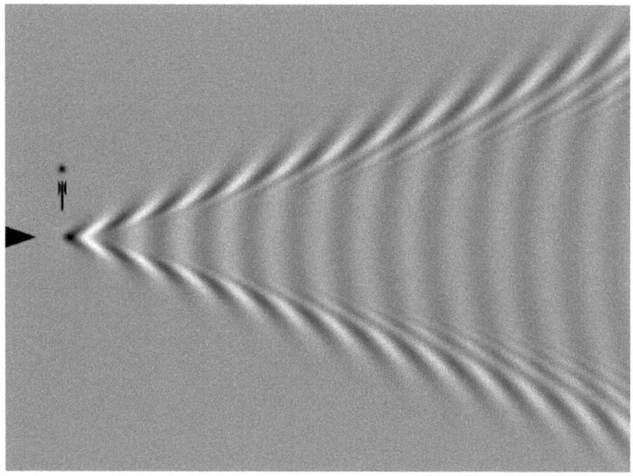

FIGURE 10.2 Simulation of a Kelvin wake. (*Image credit: Wikipedia Commons. Used with permission under CC BY-SA 4.0.*)

10.2 Ocean Remote Sensing

Synthetic aperture radar is singularly suited for the remote sensing of Earth's oceans. Much of the water masses on the planet are cloud covered, making optical imaging difficult or impossible. SAR penetrates the cloud cover and can measure the wave structure at the ocean's surface, as well as providing prominent visual features for surfactants and ships.

10.2.1 Wakes

Wakes are disturbances on the surface of water and, in the case of internal waves, under the water. In fluid dynamics, a wake can be either the area of recirculating flow immediately behind a moving or stationary object (caused by viscosity), or, the wave pattern that manifests itself on the water's surface downstream of an object in motion. For example, Figure 10.2 shows a simulation of a Kelvin wake. A common place to see a Kelvin wake is behind a boat or ship as its traveling through the water; it is named for Lord Kelvin, who first described the wake mathematically. Figure 10.3 shows an optical image of a Kelvin wake with each of the components of the Kelvin wake plainly visible (A) bow wave, (B) stern wave, (C) transverse wave, (D) turbulent wake, and (E) turbulence region adjacent to the ship.[4]

Waves and wakes manifesting themselves in SAR imagery are highly dependent on SAR geometry, particularly the grazing angle.[5] As the grazing angle gets shallower (smaller grazing angles), the wake and wave features become less and less pronounced. There is a point of diminishing returns, where the wake and/or wave will no longer be visible. Figure 10.4 shows some examples of Kelvin wakes from large vessels in Tokyo

[4] C. R. Jackson and J. R. Apel, eds., *Synthetic Aperture Radar Marine User's Manual*, U.S. Department of Commerce and the National Oceanic and Atmospheric Administration, Chapter 12, September 2004.

[5] Ibid.

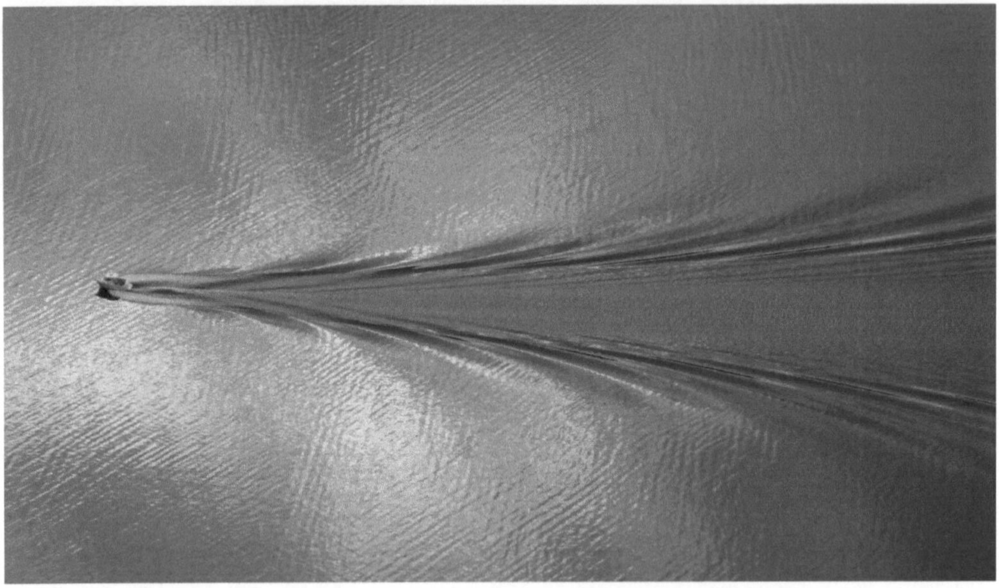

FIGURE 10.3 An optical photograph of a small boat creating a Kelvin wake. (*Image credit: Wikipedia Commons. Used with permission under CC BY-SA 3.0.*)

Bay, off the coast of Japan, as imaged by ICEYE. The grazing angle in this image is relatively steep, which makes the transverse waves within the Kelvin wake visible. As the grazing angle gets smaller, the returns from the transverse waves are attenuated, making it more difficult to see the Kelvin wake.

Internal waves, on the other hand, are waves induced by gravity that oscillate within the water, rather than on the surface. However, the internal wave can interact with the ocean's surface to produce a visible signature in synthetic aperture radar imagery. Jackson et al.[6] outline the theory behind the visualization of internal waves in SAR imagery, as well as several examples. One such example is in their Figure 1, shown in Figure 10.5, which illustrates several examples of internal waves in the Gulf of California with L-band SAR. The imagery, taken in September 1978, shows distinctive features associated with internal waves, the waves appearing as light and dark bands in the imagery.

10.2.2 Ice Monitoring

Synthetic aperture radar, particularly space-based SAR, is used extensively in the monitoring of ice. Multipolarization and multifrequency data from satellite SAR systems are used for estimating various sea ice parameters, such as ice type, distribution, and drift.[7]

[6] C. Jackson, J. da Silva, G. Jeans, W. Alpers, and M. Caruso, "Nonlinear Internal Waves in Synthetic Aperture Radar Imagery," *Oceanography*, Vol. 26, No. 2, June 2013.

[7] W. Dierking, "Sea Ice Monitoring by Synthetic Aperture Radar," *Oceanography*, Vol. 26, No. 2, June 2013, doi: 10.5670/oceanog.2013.33.

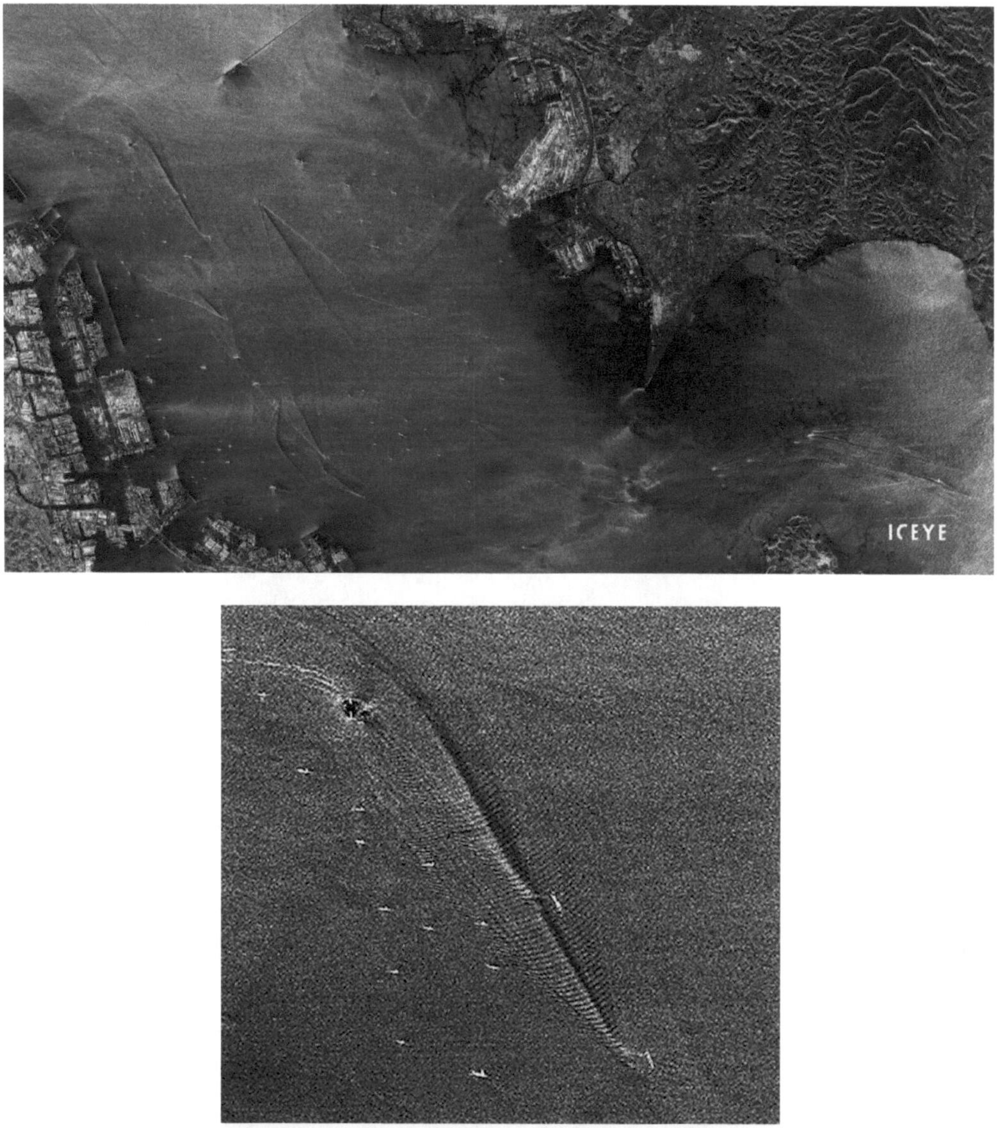

Figure 10.4 ICEYE image of Tokyo Bay, Japan. The range direction is along the rows of the image, while the azimuth direction is along the columns. The top image is the full image, while the bottom image is a zoom in on two maritime vessels heading into the bay (as seen on the left-hand side of the full image). (*Reprinted with permission.*)

SAR data can be used for creating maps of sea ice conditions. These maps are used primarily by marine traffic. The data also serves as a way to validate models that estimate and predict the kinematics and dynamics of ice. The validation of these models requires data that has resolutions of 10 kilometers or better. Figure 10.6 shows a derived ice monitoring product, offered by the National Oceanic and Atmospheric Administration

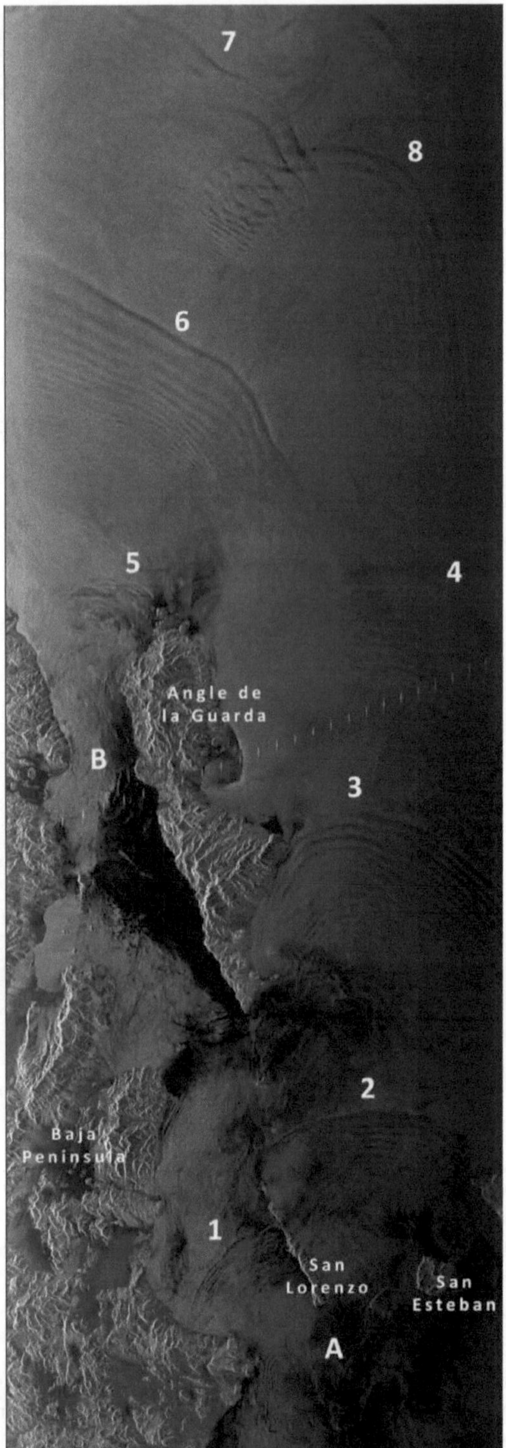

FIGURE 10.5 From Jackson et al. (2013), a Seasat L-band SAR image of the Gulf of California, showing several different types of internal waves. The dots across the center of the images are due to system calibration pulses. The area of the images is approximately 100 × 270 kilometers. (*Public domain. Used with permission*.)

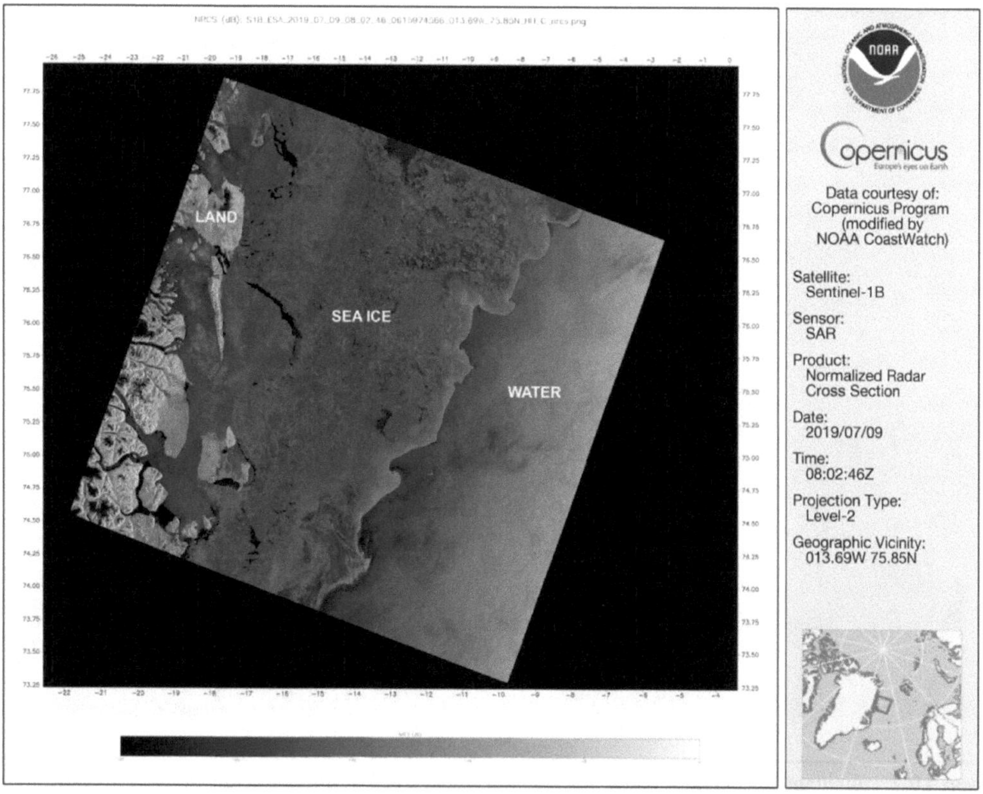

FIGURE 10.6 An example of a sea ice product from the National Oceanic and Atmospheric Administration. The data, taken in July 2019, was acquired by the Sentinel-1 satellite. The ice being monitored is off the coast of Greenland. [*Image credit: NOAA (public domain). Used with permission.*]

(NOAA). This product was created from the European Space Agency satellite, Sentinel-1. It illustrates where sea ice is present with relation to the coast line. This is critical for maintaining the safety of shipping traffic, as well as for monitoring climate.

Sea ice is a very dynamic geological structure. Changes can happen within hours. As such, it's critical to provide a temporal cadence that can effectively measure the change in sea ice as it occurs daily. Leveraging several different commercial and scientific systems is need to effectively model the change. ICEYE, the first commercial company to demonstrate synthetic aperture radar in a small sat, is planning on launching approximately 18 operational X-band SAR satellites. This will give an unprecedented temporal cadence that is not currently available (at the time of this writing). ICEYE offers both an ice monitoring product and an iceberg tracking product, both for maritime domain awareness and safety.

10.2.3 Oil Spills

Synthetic aperture radar has been used since the 1990s to detect and characterize oil spills in Earth's oceans. For example, the National Oceanic and Atmospheric Administration

Figure 10.7 An image of the east coast of Mauritius where the MV Wakashio ran aground on July 25, 2020. Image was taken on August 22, 2020 from the Sentinel-1 C-band SAR satellite. The oil is being pulled westward into the bay. (*Used with permission*.)

(NOAA) used SAR data on a near-daily basis during the Deepwater Horizon disaster in 2010. Again, because SAR does not require cloudless skies or daylight to operate, unlike optical imaging, the modality provides a singular advantage when monitoring oil spills in water. SAR can detect very thin surface oil slicks.[8] It does this because the oil attenuates the surface capillary waves—commonly known as ripples—which in turn creates dark spots in the SAR imagery.

Figure 10.7 shows an oil spill from the MV Wakashio, a Japanese tanker that ran aground off the east coast of Mauritius in July 2020. The imagery is from the Sentinel-1 satellite and is a IW ground medium resolution product, where the pixel spacing is 40 meters in both range and azimuth. The polarization is VV and the data was taken on August 22, 2020. The bright point target off the coast is the vessel itself. From the vessel are dark strands that run into the bay near Mahebourg, Mauritius. These dark strands are the oil slick that sits on the surface of the ocean. The oil appears to be flowing toward the bay and its beaches.

[8] J. V. Holmes, G. Graettinger, and I. R. MacDonald, "Remote Sensing of Oil Slicks for the Deepwater Horizon Damage Assessment," Chapter 17 in *Oil Spill Science and Technology*, Elsevier, 2017, pp. 889–923.

FIGURE 10.8 The Mauritius oil spill as seen from the electro-optical satellite Sentinel-3. Note the presence of clouds that are not present in the Sentinel-1 SAR data. This data was taken on August 11, 2020. After some study, the oil slicks are visible in the electro-optical imagery, but are not nearly as detectable visually as they are in the SAR data. (*Used with permission under CC BY-SA 3.0.*)

Figure 10.8 shows a Sentinel-3 electro-optical (EO) image of the same area in Mauritius, taken on August 11, 2020. The original RGB image was converted to grayscale for publication. The oil slick is harder to see in the EO than in the SAR imagery. This is because the primary observable in the EO imagery is the color change between the oil and water, which does not have a high contrast, rather than the high contrast provided by the dampening of the capillary waves in the SAR image. This dampening reduces the ocean clutter, making the oil slick darker than the surrounding water.

10.3 Deep Learning

There has been a good deal of work examining synthetic aperture radar and deep learning for ATR. An excellent survey of deep learning for remote sensing can be found in Zhu et al.,[9] which includes a survey of techniques used in synthetic aperture radar. The types of machine learning used are diverse, spanning from support vector machines with identified features to deep learning with convolutional neural networks (CNNs). In most cases, however, the information comes from the magnitude image, rather than the phase. This could potentially limit the useful information for phase-varying targets, such as moving objects. Additionally, this requires processing the raw radar pulses to images using computationally expensive algorithms such as Omega-K or the Polar Formatting Algorithm (PFA). The creation of the image product generates additionally computational overhead that can limit the ability to perform real-time ATR.

Deep learning is a relatively new field of machine learning. Leveraging the work done in the 1990s in neural networks, deep learning uses multiple layers to converge on a set of features that can then be used in a discrimination algorithm, such as a fully connected neural network or support vector machine, to do classification. In the case of synthetic aperture radar, the data is given to the first set of layers and, through the mathematical operation of convolution, inherent characteristics of the image are grouped and ranked. The grouping and ranking happens are part of the training of the overall network. This training, if done properly, not only selects the best features for the class of objects being considered, but also the decision boundary in the discriminator to make the best decision about which class the object belongs to. These networks that employ convolution to develop inherent features for the objects are called *convolutional neural networks*.

For example, Soldin[10] in 2018 applied the ResNet-18 architecture to the MSTAR data set. This data set is a staple among machine learning algorithm developers as it is freely available and accurately labeled. There are thousands of these target chips, making for a rich training and test set. The data, taken in the late 1990s by Sandia National Laboratory, consists of military vehicles imaged with spotlight SAR in a 1-foot resolution mode. The images were taken a various grazing angles and aspect angles. Figure 10.9 shows some of the target chips, along with digital photographs of the vehicle for reference.

Using 3671 different chips for training and 3203 different chips for testing, Soldin was able to achieve classification accuracies of anywhere from 86% to 100%, depending on the target type. He also examined the use of subapertures to create a three-band image for classification. This produced slightly better results. The overall accuracy for the ResNet-18 network was about 81%, when training from scratch and fine-tuned. The overall accuracy of the subaperture model was about 83%.

It is worth considering that the formation of the image provides no added information to the synthetic aperture radar data. The image is simply arranged in a way that is

---

[9] X. X. Zhu et al., "Deep Learning in Remote Sensing: A Comprehensive Review and List of Resources," *IEEE Geoscience and Remote Sensing Magazine*, Vol. 5, No. 4, pp. 8–36, December 2017, doi: 10.1109/MGRS.2017.2762307.

[10] R. J. Soldin, "SAR Target Recognition with Deep Learning," in 2018 IEEE Applied Imagery Pattern Recognition Workshop (AIPR), Washington, DC, October 2018, pp. 1–8, doi: 10.1109/AIPR.2018.8707419.

FIGURE 10.9 Some examples of the MSTAR target chips, freely available from Air Force Research Laboratory (AFRL). The MSTAR target chips are shown on the right, with a photograph of the machine imaged on the left. [*Image credit: AFRL/Sandia (public domain). Used with permission.*]

visually useful to a human. In fact, when making the magnitude image for human visualization, information is lost. The phase is discarded. Most deep learning efforts work in the magnitude image domain, likely because canned algorithms for image recognition already exist and can be easily applied to SAR magnitude image data. There are some notable exceptions to this. Alver et al.[11] examined the use of the complex-valued phase

[11] M. B. Alver, S. Atito, and M. Çetin, "SAR ATR in the Phase History Domain Using Deep Convolutional Neural Networks," in Image and Signal Processing for Remote Sensing XXIV, Berlin, Germany, October 2018, p. 41, doi: 10.1117/12.2325365.

histories to a deep convolutional neural network. They achieved an overall accuracy of 84% across 10 classes of targets, using the MSTAR data set.

Kazemi, Yonel, and Yazici[12] employed a recurrent neural network to learn what targets look like using single radar pulses across the synthetic aperture. In doing so, they were able to demonstrate that it is possible to recognize targets in a single range profile or a small set of range profiles. Using synthetic data, they were able to achieve a 97% to 98% accuracy, depending on the dictionary size used for the recurrent network.

[12] S. Kazemi, B. Yonel, and B. Yazici, "Deep Learning for Direct Automatic Target Recognition from SAR Data," in 2019 IEEE Radar Conference (RadarConf), Boston, MA, April 2019, pp. 1–6, doi: 10.1109/RADAR.2019.8835492.

CHAPTER 11
Interferometry

Synthetic aperture radar (SAR) is a coherent imaging modality. As such, it preserves the phase of the scene that's being imaged. This phase can be exploited for a multitude of applications, which include terrain mapping [e.g., the creation of digital elevation models (DEMs)], coherent change detection, and changes in terrain (e.g., the shifting of terrain after an earthquake). This chapter will focus on the basics of interferometry and how you can measure small changes in phase between two images.

11.1 The Basics

The thought of measuring small changes in phase is often confusing to many SAR practitioners. If you view the phase in an image, SAR phase looks random to the human eye. It is difficult or impossible to see any structure that could be exploited in a phase image. However, this is not the case. SAR phase, absent of significant noise sources, is consistent between two images. Thus, if you took two images at the exact same geometry at two different times, you would see that the phases would cancel out if you subtract them from one another. This assumes, of course, that the scene didn't change between the time the first image was taken and the second image was taken. If the phases do not subtract out, then there was a change in the terrain and this can be measured to a quarter of the wavelength of the radar. This is called *coherent change detection*.

Recall that we have assumed up to now that a scene was static (nothing moves in it) and flat (there is no terrain elevation). In the case of a dynamic scene, such as a moving car, this assumption is violated and creates a unique signature in the image. We will examine this in Chapter 12. If there is terrain elevation, this also causes a unique set of conditions in the imagery. This effect, typically called "layover," will be described here. We can take two images of the same scene, but at slightly different viewing angles, and measure the difference in phase between them. This difference in phase corresponds directly to the difference in elevation. It is through this that digital elevation models (DEMs) can be created. These techniques are called interferometric synthetic aperture radar and are often abbreviated IFSAR or InSAR.

There are several excellent papers and tutorials on interferometric SAR. One, by Fornaro, Lombardini, Pauciullo, Reale, and Viviani,[1] is an excellent survey of how

---

[1] G. Fornaro, F. Lombardini, A. Pauciullo, D. Reale, and F. Viviani, "Tomographic Processing of Interferometric SAR Data: Developments, Applications, and Future Research Perspectives," *IEEE Signal Processing Magazine*, Vol. 31, No. 4, pp. 41–50, July 2014, doi: 10.1109/MSP.2014.2312073.

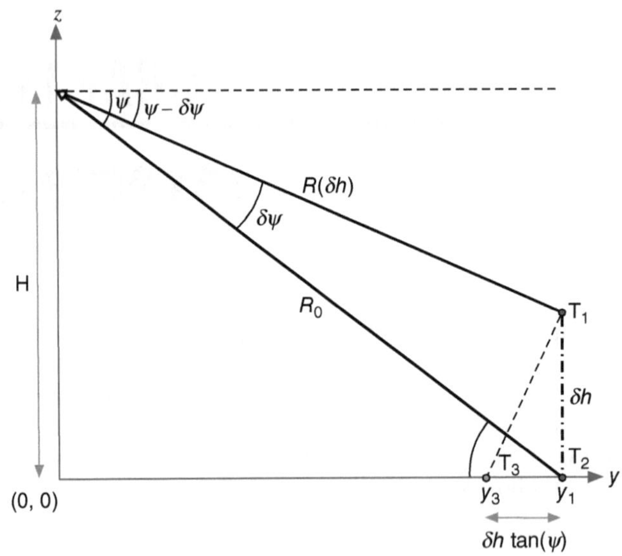

FIGURE 11.1 The basic geometric model for interferometric SAR. This model also illustrates how layover works in SAR.

tomographic processing (e.g., backprojection) can be used to create elevation maps. They also discuss how interferometry can be used to cluster concentrated scatterers; they describe how this can be used for estimating the stability of buildings using this method.

Another excellent tutorial on interferometric SAR comes from Richards.[2] Richards focuses on the basics of interferometry with its application to the creation of DEMs. This section will summarize Richard's excellent tutorial.

Figure 11.1 shows the basic geometry for considering how height plays a factor on SAR phase history. Richards showed that the phase of a scatterer has the form

$$\Phi \equiv \Phi_A + \Phi_\rho - \frac{4\pi}{\lambda} R_0 \tag{11.1}$$

where A is the complex-valued constant that considers the original transmitted amplitude, all the radar range equation influences other than the radar cross section, and the radar receiver's complex gain, and ρ is the complex reflectivity of the scatterer (ρ is proportional to $\sqrt{\sigma}$, which is the scatterer's radar cross section).

Without loss of generality, this phase corresponds to the phase of T_2 in Figure 11.1. This means that it doesn't matter if quadrature demodulation and matched filtering or dechirp-on-receive was used to produce the processed radar data. Now consider the scatterer at T_1 in our geometry figure. The origin of this geometry has been labeled as $(y, z) = (0, 0)$, but is really $(y, z) = (0, h_{ref})$, where h_{ref} is some elevation that holds for the

[2] M. A. Richards, "A Beginner's Guide to Interferometric SAR Concepts and Signal Processing," *IEEE A&E Systems Magazine*, Vol. 21, No. 6, p. 25, 2006.

entire scene. For simplicity, we will continue to consider $h_{\text{ref}} = 0$. The height of T_1 is thus δh. This implies that the range to T_1 is

$$R(\delta h) = \sqrt{y_1^2 + (H - \delta h)^2} = \frac{H - \delta h}{\sin(\Psi - \delta\Psi)} \tag{11.2}$$

It is clear, both from the illustration and from Equation (11.2), that T_2 is closer to the radar than T_1. This reduction in range results in T_1's echo arriving *before* T_2. This early arrival means that the target will "layover" in the scene closer to the radar than T_2. A tall object in a SAR image will always point toward, and in the direction of, the radar. The distance of this layover from the sensor's ground track is

$$y_3 = y_1 - \delta h \tan(\Psi) \tag{11.3}$$

In addition to laying over in the scene, the delayed echo will also cause a change in the phase of the scatterer. This can be quantified by taking the derivative with respect to δh. Richards demonstrated that this results in

$$\delta R = -\left[\frac{H - \delta h}{R(\delta h)}\right]\delta h \tag{11.4}$$

The effect on the range of a deviation in T_1's height from the reference plane h_{ref} can be found by examining $\delta h = 0$. This evaluation produces

$$\delta R = -\left(\frac{H}{R_0}\right)\delta h = -\delta h \sin(\Psi) \tag{11.5}$$

where Ψ is the grazing angle and R_0 is the range to T_2 (essentially the return from a scatterer at the ground plane). This change in range is proportional to the change in round-trip time delay,

$$\frac{2\delta R}{c} = \frac{2\delta h \sin(\Psi)}{c} \tag{11.6}$$

Thus, the change in phase due to a received echo at some height δh will be

$$\delta\Psi = -\left(\frac{4\pi}{\lambda}\right)\delta R = \left(\frac{4\pi}{\lambda}\right)\delta h \sin(\Psi) \tag{11.7}$$

Now let's consider how height effects phase and what we will defined as the interferometric phase difference (IPD). Let's imagine that there are two SAR images taken at the sensor location in Figure 11.1. We will call one $f(x, y)$ and the other $g(x, y)$, as Richards did. In interferometric SAR (IFSAR), there are two ways to acquire these images:

- One-pass interferometry: f and g are two different phase apertures on the same aircraft or spacecraft; they image the scene at the same time and are separated by B meters.

- Two-pass interferometry: f and g are two different images taken (typically) with the same SAR at different times; they are also separated by B meters and the assumption is that their imaging geometries were identical but their locations in the range direction are different.

If the two images are separated in the range direction by B meters, then they are separated in the slant plane by $B\cos(\Psi)$, again where Ψ is the grazing angle. An example of a two-pass system is the Sentinel-1 SAR. This system transmits and receives using the same antenna, which is a single-phase center or aperture. IFSAR can be performed on two different images", striking the "one" that were taken of the same patch of ground, just at different positions on the two passes. This results in a difference in range of $2\Delta R$. A one-pass system would have a difference of path length at the two phase centers of ΔR. Let's continue with the two-pass case. Both images, $f(x, y)$ and $g(x, y)$ are complex-valued images containing both magnitude and phase of the scene. The difference in phase, called the IPD, is defined as

$$\phi_{fg}(x, y) \equiv \phi_f(x, y) - \phi_g(x, y) \approx -\frac{4\pi}{\lambda} B\cos(\Psi) \tag{11.8}$$

Richards then defines the "flat-Earth" phase difference, ϕ_{fg}^{FE}:

$$\begin{aligned} \phi_{fg}^{\mathrm{FE}}(x, y) &\approx -\frac{4\pi}{\lambda} B\cos\left[\Psi(y)\right] \\ &= -\frac{4\pi B}{\lambda}\left(\frac{y}{R_0}\right) \\ &= -\frac{4\pi B y}{\lambda\sqrt{H^2 + y^2}} \\ &= -\frac{4\pi B}{\lambda\sqrt{1 + (H/y)^2}} \end{aligned} \tag{11.9}$$

Richards points out that this phase difference is dependent on the target's ground range in the y direction, but not on its azimuth coordinate x. Naturally, as the equation suggests, there is a phase difference in range when looking from two different locations (in range). However, when you have two different azimuth locations, the same range geometry is presented and thus, f and g have the same IPDs.

This flat-Earth phase difference is removed during interferometric processing to produce

$$\phi'_{fg} \equiv \phi_{fg} - \phi_{fg}^{\mathrm{FE}} \tag{11.10}$$

This normalizes the phase to the reference height h_{ref}. Any additional variations in the IPD, ϕ'_{fg} will be due to change in elevation. As can be seen in Figure 11.1, the grazing angle between T_1 and the sensor changes and is represented by $\delta\Psi$. The subsequent change in phase can be found by taking the derivative of Equation (11.8) with respect to Ψ:

$$\frac{d\phi'_{fg}}{d\Psi} = \frac{4\pi}{\lambda} B\sin(\Psi) \tag{11.11}$$

This leads then to the definition of the change in grazing angle between the elevated scatterer and the ground plane,

$$\delta\Psi = \frac{\lambda}{4\pi B \sin(\Psi)} \delta\Phi'_{fg} \tag{11.12}$$

Richards goes on to derive the relationship between the change of height in the modified IPD by using a series expansion for the tangent function and making the reasonable assumption that $\delta\Psi$ is relatively small. With this, he asserts that

$$\delta h(x,y) = \frac{\lambda H \cot(\Psi)}{4\pi B \sin(\Psi)} \delta\phi'_{fg}(x,y) \tag{11.13}$$

This equation is the foundation of IFSAR. Richards goes on to introduce the interferometric scale factor α_{IF} and states that $\delta h(x,y) \equiv \alpha_{\text{IF}}\delta\phi'_{fg}$. He goes on to note that the scale factor is defined as

$$\alpha_{\text{IF}} = \frac{\lambda H \cot(\Psi)}{4\pi B \sin(\Psi)} \tag{11.14}$$

11.2 Terrain Estimation

Using Equation (11.13), an estimate of the terrain can be found. Consider a reference point (x_0, y_0) in the image, where the height is well known or at least easily defined, such that $h(x_0, y_0) = h_0$. An estimate of the height of a given pixel can be found by comparing its IPD to that of an adjacent pixel, such that,

$$h(x_1, y_1) - h(x_2, y_2) \approx \alpha_{\text{IF}} \left[\phi'_{fg}(x_1, y_1) - \phi'_{fg}(x_2, y_2)\right] \tag{11.15}$$

Thus, the terrain map can be calculated by

$$\hat{h}(x,y) = h_0 + \alpha_{\text{IF}} \left[\phi'_{fg}(x,y) - \phi'_{fg}(x_0, y_0)\right] \tag{11.16}$$

As an example of this, consider the Shuttle Radar Topography Mission (SRTM). This mission sought to create digital elevation models for nearly the entire planet. The SAR sensor used was based on the SIR-C/X-SAR sensor. The SRTM was used in one-pass interferometric mode, having two antennas installed. The antennas were separated by a 60-meter mast, with one antenna located in the shuttle's bay and the other at the end of the mast. The mission flew for 11 days in 1995 and produced elevation models that are arranged in tiles, each covering 1° of latitude and 1° of longitude. The cross-range resolution of the data was approximately 30 meters, while the post spacing was 30 meters. The relative vertical accuracy was, on average, between 15 and 50 meters. The system used HH polarization to image. An example of a SRTM terrain model can be seen in Figure 11.2. This model of central Nevada was created in 1995. The 30 meter post spacing provides for a detailed map where the topography of the mountain ranges in Nevada is well characterized.

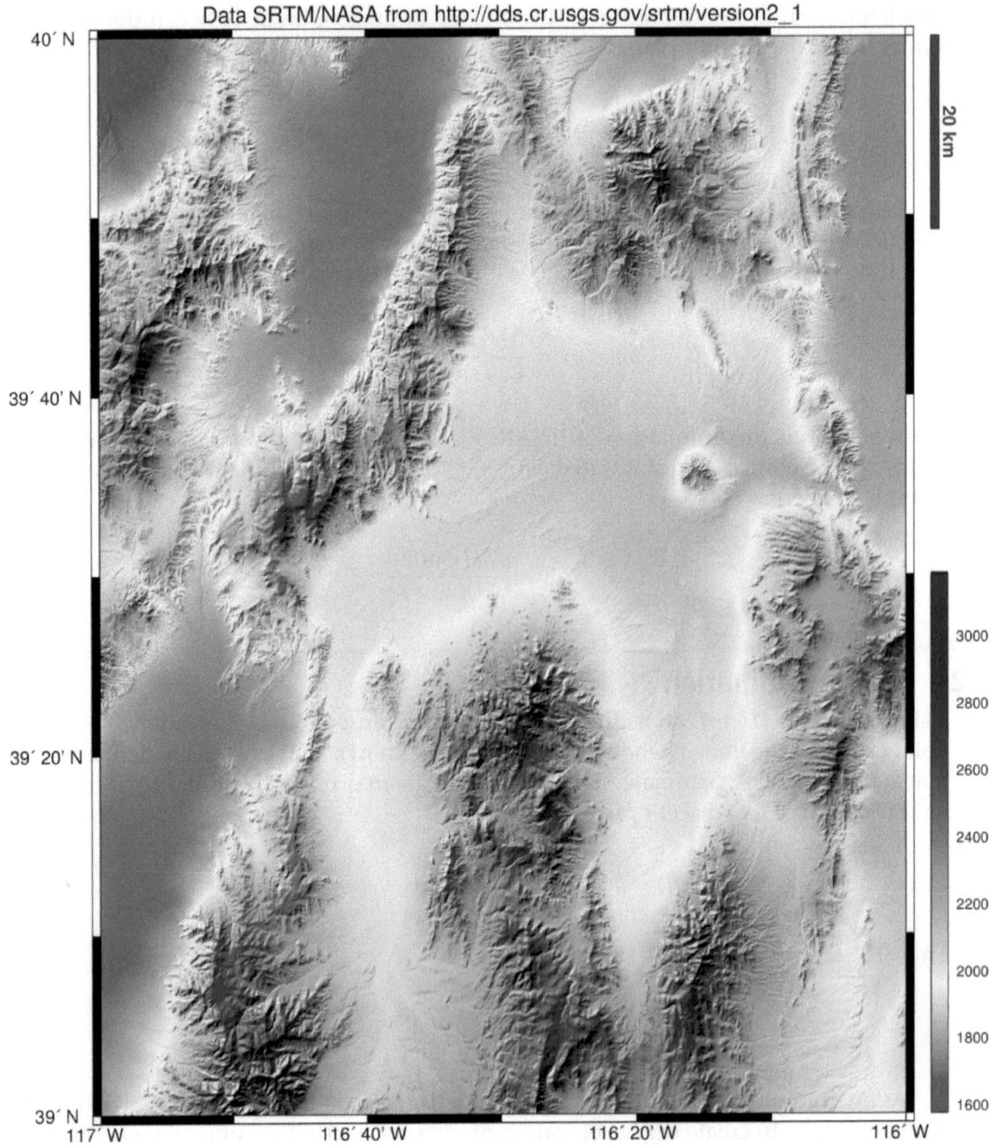

FIGURE 11.2 An example of a terrain model of central Nevada produced by SRTM in 1995. (*Image by François Beauducel. Permission to reprint under CC BY-SA 3.0.*)

11.3 Coherent Change Detection

Jakowatz et al.[3] discuss a two-pass interferometric method called *coherent change detection* (CCD). The two images used in CCD are taken at two different times, likely hours or days

[3] C. V. Jakowatz, D. E. Wahl, P. H. Eichel, D. C. Ghiglia, and P. A. Thompson, *Spotlight-Mode Synthetic Aperture Radar: A Signal Processing Approach*, Kluwer Academic Publishers, Boston, MA, 1996.

apart from one another. The assumption now is that the terrain height has not changed, but that the complex reflectivity ρ has. This change is commonly due to seemingly benign effects, such as disturbance of the ground by moving vehicles or persons walking. However, CCD can also measure big changes, such as the demolition and removal of a building. The small changes can be measured to a fraction of the radar's wavelength. As Richards[4] points out, if there is no difference in the reflectivity of a given pixel, (x, y) between the two time periods, computing a normalized correlation coefficient of the two measurements of reflectivity maps, $\rho_1(x, y)$ and $\rho_2(x, y)$, should produce a value approximately equal to 1.0. That is, the reflectivity between the two pixels is highly correlated and not much has changed. The bigger the change in reflectivity in the pixel, the closer to zero the value of the correlation becomes.

This can be seen in the example in Figure 11.3. ImSAR took two different images with their Ku band nano-SAR pod (NSP). The first image is the reference image, which is shown at the top of the figure. The second image, taken some hours later after a vehicle had passed through, is shown in the middle of the figure. The CCD product, which is the correlation between the two reflectivities, is shown at the bottom. The tracks where the vehicle had driven are plainly visible in the CCD product. Trees and shrubs are also visible; they likely show as different because of reflectivity changes due to wind.

11.4 Earthquakes

Turning the problem around a bit, IFSAR can be used to measure the *height change* over time. That is, a two-pass configuration can be used, again, where the images are taken hours or days apart, and a map of the terrain change can be created. This map is useful for all sorts of applications, such as tectonic activity,[5] tunnel detection[6] (particularly for Earth deformation for safety and civic development), and other changes to the Earth's surface due to human activity.

Figure 11.4 shows an example of an IFSAR product that measures the change in terrain height after a series of earthquakes in Southern California. These swarms of earthquakes occurred in 2019 near Ridgecrest, CA and Trona, CA with two major events occurring on July 4 and July 5, 2019, respectively. The magnitudes of the two major events were 6.4 on July 4 and 7.1 on July 5. The figure shows the fringe pattern of the change in phase between two images taken with the ALOS-2 satellite, operated by the Japan Aerospace Exploration Agency (JAXA). The reference image was taken on April 16, 2018. The second image was taken on July 8, 2019. Each gray color cycle represents 4.8 inches (12 centimeters) of ground displacement.[7]

---

[4] M. A. Richards, "A Beginner's Guide to Interferometric SAR Concepts and Signal Processing," *IEEE A&E Systems Magazine,* Vol. 21, No. 6, p. 25, 2006.

[5] T. J. Wright, Z. Lu, and C. Wicks, "Constraining the Slip Distribution and Fault Geometry of the M_w 7.9, 3 November 2002, Denali Fault Earthquake with Interferometric Synthetic Aperture Radar and Global Positioning System Data," *Bulletin of the Seismological Society of America,* Vol. 94, No. 6B, 2004.

[6] Z. Wang, D. Perissin, and H. Lin, "Subway Tunnels Identification Through Cosmo-SkyMed PSInSAR Analysis in Shanghai," *2011 IEEE International Geoscience and Remote Sensing Symposium,* Vancouver, BC, Canada, 24–29 July 2011.

[7] Details at `https://www.jpl.nasa.gov/spaceimages/details.php?id=PIA23150`

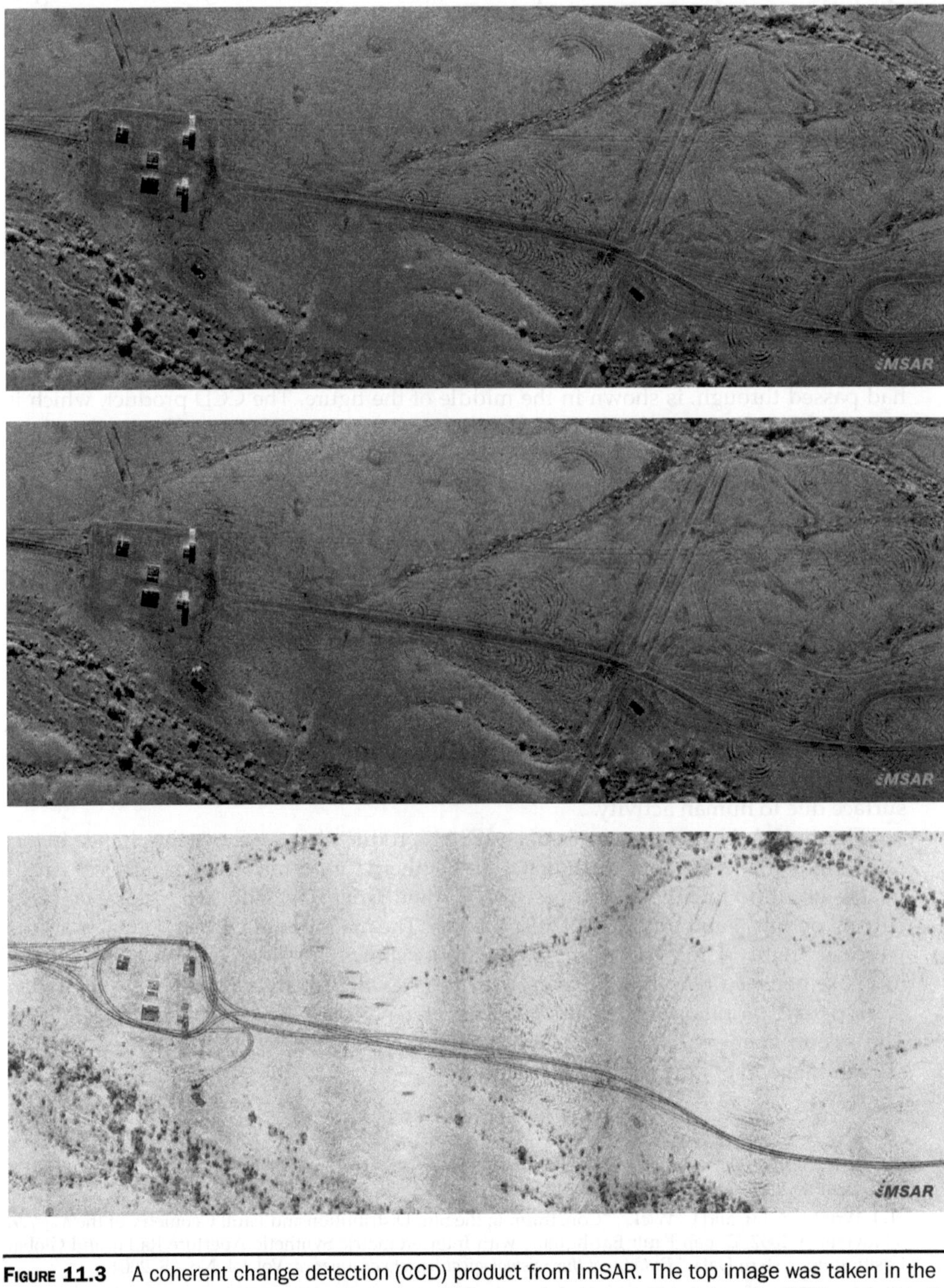

Figure 11.3 A coherent change detection (CCD) product from ImSAR. The top image was taken in the first pass and the second image, shown in the middle, was taken sometime later. The CCD product is the bottom image, where white pixels correspond to little to no change and black pixels correspond to significant change. (*Used with permission.*)

FIGURE 11.4 Interferometric SAR (IFSAR) map showing the surface displacement caused by major earthquakes in Southern California, near Trona, CA and Ridgecrest, CA. (*Courtesy: NASA/ JPL-Caltech. (Public domain). Used with permission.*)

The resolution of the images is about 90 meters. The line across the fringes likely indicates possible surface ruptures. The product was created by the NASA Jet Propulsion Laboratory and California Institute of Technology's Advanced Rapid Imaging and Analysis (ARIA) team.

CHAPTER 12

Moving Objects in Synthetic Aperture Radar

The basic mathematical model of synthetic aperture radar (SAR) assumes that the imaging scene is flat and that nothing moves. In the previous chapter, we learned what happens with phase when the scene isn't flat. You will recall that interferometry can be used to measure the change in phase and that change in phase is directly proportional to elevation.

When an object, such as a vehicle, moves in the scene during imaging, two things can happen: the target will displace or shift away from its original location, that shift is always in the azimuth direction in the image, and, the target will smear along the azimuth direction. This is often called the "train off the tracks" effect.[1] This effect can be seen when a train is imaged; the train is offset from the tracks, smeared, and its radar shadow remains next to the track. The degree of displacement and smearing is directly related to the moving object's kinematics. This chapter describes the mathematical models behind moving objects in SAR imagery, as well as exploring the difficulty of measuring the kinematics from the displacement and smear.

Figure 12.1 shows a literal example of the "train off the track" phenomenon. This figure has a Sentinel-1 image taken on August 16, 2020. The image covers a very large area of Southern California, including Banning's Pass, the Coachella Valley (including Palm Springs, CA), the Salton Sea, and the Imperial Valley. In the right-hand corner, there is a small dark box. This box encompasses a railway near Andrade, CA, which is the southeast most corner of California, very near Yuma, AZ. The imagery in the box has been expanded in the upper right corner to show the track, which starts at the left of the box and runs all the way to the right. Just below the track is a bright line running from left to right, about a quarter of the way in from the left edge of the image to a quarter of the way from the right edge of the image. This is a long train running along the track. The train is moving almost entirely in the range direction, so it is displacing in azimuth. There is very little smearing, indicating that there is very little motion in the azimuth direction. We will outline the theory of this in the next section.

---

[1] C. L. Rufenach, R. A. Shuchman, and D. R. Lyzenga, "Interpretation of Synthetic Aperture Radar Measurements of Ocean Currents," *Journal of Geophysical Research*, Vol. 88, No. C3, p. 1867, 1983, doi: 10.1029/JC088iC03p01867.

FIGURE 12.1 A Sentinel-1 image in high-resolution interferometric wide mode (10-meter resolution) of Southern California. The dark box in the lower right corner has been zoomed in the upper right corner. This is an area near Andrade, CA. where there is a railway. The "train off the track" phenomenon can be plainly seen, where the long train has displaced in the azimuth direction below the track. (*Public domain. Used with permission.*)

12.1 Mathematical Model

There are several different approaches that we could take in examining the mathematics behind moving objects in SAR. Raney[2] examined the change in range by looking at the instantaneous position of the object versus the radar and expanded range using a Maclaurin series. Fienup[3] examined the range profile of the moving object using a more generalized Taylor series expansion. In each case, the models require making an approximation. As we will see in this chapter, these approximations can fool someone into believing that you can derive the moving object's kinematics from a single-channel SAR image. That is, a SAR with only one receive antenna phase center.

[2] R. Raney, "Synthetic Aperture Imaging Radar and Moving Targets," *IEEE Transactions on Aerospace and Electronic Systems*, Vol. AES-7, No. 3, pp. 499–505, May 1971, doi: 10.1109/TAES.1971.310292.

[3] J. R. Fienup, "Detecting Moving Targets in SAR Imagery by Focusing," *IEEE Transactions on Aerospace and Electronic System*, Vol. 37, No. 3, pp. 794–809, July 2001, doi: 10.1109/7.953237.

To derive the model for a moving object in synthetic aperture radar, consider a moving object with a position in the scene of

$$(x, y) = (x_0 + v_r\eta + \frac{1}{2}a_r\eta^2, y_0 + v_a\eta) \tag{12.1}$$

where $-T/2 \leq \eta \leq T/2$. Note that acceleration is only considered in the range direction, not in the azimuth direction. Fienup explains that this is a simplifying assumption. Raney, however, included acceleration in both range and azimuth and demonstrates that only range acceleration plays a factor. With this, we can show that the change in phase over slow time due to the moving object is

$$\Phi(f', \eta) = \frac{4\pi}{\lambda}\left(r(\eta) - \sqrt{R_0^2 + (v_p\eta)^2}\right) \tag{12.2}$$

where $r(\eta) = \sqrt{(R_0 + x_0 + v_r\eta + \frac{1}{2}a_r\eta^2)^2 + (y_0 + v_a\eta + v_p\eta)^2}$ and $f' = (f - f_0)$. Using a Taylor series expansion, as Fienup demonstrated, phase can be described by the following terms:

$$\begin{aligned}\Phi(f', \eta) = \frac{4\pi}{\lambda}\Bigg[&x_0 + x_0\frac{f'}{f_0} - y_0\frac{v_p\eta}{R_0} \\ &+ v_r\eta + \frac{1}{2}a_r\eta^2 - \frac{v_a}{v_p}\frac{(v_p\eta)^2}{R_0} \\ &- \frac{v_a}{v_p}\frac{f'}{f_0}\frac{(v_p\eta)^2}{R_0} + \cdots\Bigg]\end{aligned} \tag{12.3}$$

The first three terms of Equation (12.3) correspond to constant phase, the range position (x_0), and the azimuth position (y_0), respectively. The fourth term contributes to the displacement or shift in azimuth. You will note that a velocity in the range direction for the moving object is solely responsible for the shift. Thus, if there is no range velocity component (i.e., the object is moving only in the azimuth direction), there will be no shift.

The fifth and sixth terms $\frac{4\pi}{\lambda}\left(\frac{1}{2}a_r\eta^2 + \frac{v_a v_p\eta^2}{R_0}\right)$ contribute to the smearing of the moving object in the azimuth direction. These terms indicate that range acceleration and azimuth velocity components of the moving object are responsible for smear. The overall phase term is quadratic. You will recall from Chapter 7 that quadratic phase error across the entire aperture defocuses an image. Here, the quadratic phase error is localized, affecting only the moving object. If the moving object has no range acceleration and has no velocity component in the azimuth direction, then there is no smear.

Finally, the final term in the Taylor series contributes to the range walk of the smear. Range walk is equivalent to range cell migration. The target's smear cross from one range cell to the next. For many space-based synthetic aperture radar systems, this term is negligible, as R_0 tends to be much larger than $v_a v_p\eta^2$.

When inspecting the magnitude image, you can often spot the moving target by its smear. For ground vehicles, such as cars and trucks, the smear is typically displaced off the road that it is actually on. This makes visual inspection of the targets relatively easy. If there is a smeared target along azimuth near, but not on, a given roadway—or a wake for a maritime vessel—then it is likely a moving object.

As you can probably deduce, because there are instances where a target won't smear (no range acceleration and no azimuth velocity), there are moving objects that will displace and look like a focused scatterer. We will describe this effect as an "ambiguous" moving object. In fact, not only is it possible for a moving object to mimic a stationary scatterer, it is also possible for a moving object to mimic a different moving object. That is, their displacement will end up in the same place in the scene and their smears will be at the same length. This ambiguity makes it impossible to derive the kinematic information of a moving object in a single-channel, single-look SAR image.[4] We'll explore this more in the next section.

Example 12.1 Consider an airborne SAR system with the following parameters:

Parameter	Variable Name	Value
Platform velocity	v_p	57.15 m/s
Aperture time	T	1.5 s
Range to scene center	R_0	4500 m
Aperture angle	$\Delta\theta_a = \frac{v_p T}{R_0}$	0.019 rad (1.09°)
Center frequency	f_0	9.6 GHz
Bandwidth	B	591 MHz
Wavelength	λ_0	0.31 m

Now consider a moving object in the scene. For simplicity, we will consider the object to be an ideal point scatterer. The scatterer is moving away with constant from the scene center, mostly in the azimuth direction. The velocity in the azimuth direction is $v_a = 13.5$ m/s, while the velocity in the range direction is $v_r = 0.19$ m/s. This means that the point scatterer is moving diagonally with respect to the flight track of the sensor, but only slightly. For the aperture time given, the distance the object moves in azimuth is $d_a = v_a T = 20.25$ m, while in range, the scatterer moves $d_r = v_r T = 0.3$ m.

Fienup[5] derives the distance that the object shifts in the image due to range velocity as $\frac{v_r R_0}{v_p}$, while the smear due to azimuth velocity is $2v_a T$. In this example, the displacement in azimuth due to the range velocity of the object is 14.96 m, while the smear in azimuth due to the azimuth velocity of the object is 40.5 m.

From the Taylor series expansion, we know that the phase induced by the moving object's range velocity is

$$\Phi_r = \frac{4\pi}{\lambda_0} v_r \eta$$

[4] R. D. Chapman, C. M. Hawes, and M. E. Nord, "Target Motion Ambiguities in Single-Aperture Synthetic Aperture Radar," *IEEE Transactions on Aerospace and Electronic System*, Vol. 46, No. 1, pp. 459–468, January 2010, doi: 10.1109/TAES.2010.5417175.

[5] J. R. Fienup, "Detecting Moving Targets in SAR Imagery by Focusing," *IEEE Transactions on Aerospace and Electronic System*, Vol. 37, No. 3, pp. 794–809, July 2001, doi: 10.1109/7.953237.

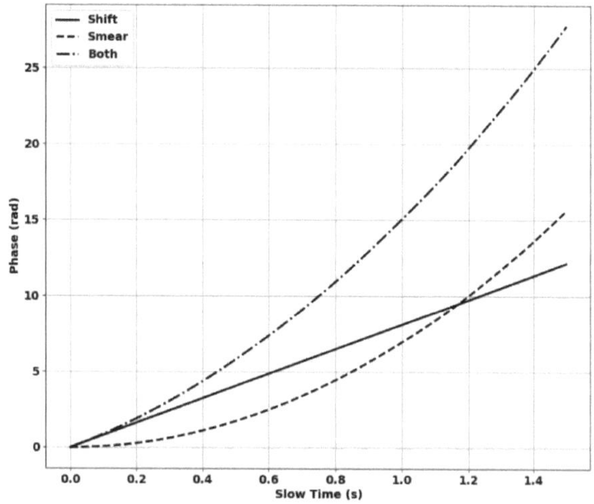

FIGURE 12.2 The phase contributions of a moving point scatterer at $v_a = 13.5$ m/s and $v_r = 0.19$ m/s.

This is a linear phase term, which equates to a shift in azimuth. We also know that the phase induced by the moving object's azimuth velocity is a quadratic term and is defined as

$$\Phi_a = \frac{4\pi}{\lambda_0}\frac{v_a v_p \eta^2}{R_0}$$

This quadratic phase term is applied to just this target (unlike the quadratic phase error that defocuses the entire image and requires autofocus) and smears the target in the azimuth direction.

Figure 12.2 shows the individual phase contributions due to v_r (Φ_r, which is the line in the graph and contributes to the object's shift in azimuth) and v_a (Φ_a, which is the quadratic in the graph and contributes to the object's smearing in azimuth). Both phases added together, as indicated by the Taylor series, contribute to the phase of that particular moving object over all of slow time.

12.2 Challenges with Moving Object Estimation and Single-Channel SAR

Consider a synthetic aperture radar that has a single transmit and receive antenna. As was mentioned previously, in this configuration, it is possible for a moving object's range profile over slow time to mimic that of a stationary object and that of another moving object. Chapman, Hawes, and Nord[6] studied this carefully and set forth a simple geometric and mathematical explanation for why this "degeneration" can occur.

[6] R. D. Chapman, C. M. Hawes, and M. E. Nord, "Target Motion Ambiguities in Single-Aperture Synthetic Aperture Radar," *IEEE Transactions on Aerospace and Electronic System*, Vol. 46, No. 1, pp. 459–468, January 2010, doi: 10.1109/TAES.2010.5417175.

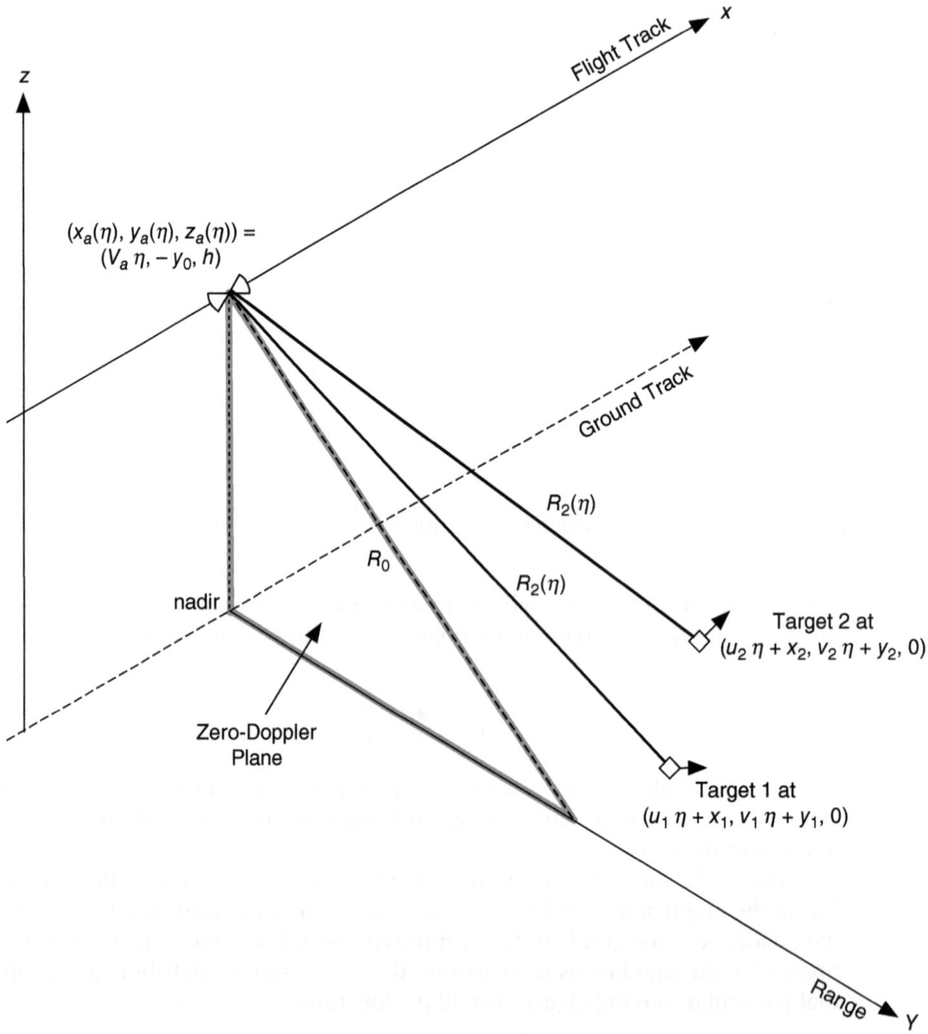

FIGURE 12.3 The geometry for a SAR system imaging two linearly moving objects.

Consider Figure 12.3, which shows a single-channel SAR system observing two objects, both with constant linear motion. Target 1 has the velocity vector (u_1, v_1) with a starting position of (x_1, y_1), while Target 2 has the velocity vector (u_2, v_2) with a starting position of (x_2, y_2). For a moving target to mimic either a stationary scatterer or another moving object, then $R_1(\eta) = R_2(\eta)$ for all of slow time during the dwell. That is

$$R_1(\eta) = R_2(\eta) \tag{12.4}$$

$$\sqrt{(V_a\eta - u_1\eta - x_1)^2 + (y_0 + v_1\eta + y_1)^2 + h^2} = \sqrt{(V_a\eta - u_2\eta - x_2)^2 + (y_0 + v_2\eta + y_2)^2 + h^2} \tag{12.5}$$

If R_1 and R_2 are equal during the entire dwell, then the two objects have the same range profile, making the two objects indistinguishable. There are two cases in particular:

- A moving object becomes ambiguous with a stationary target; for example, Target 1 moves in such a way that it mimics a stationary Target 2's range profile $((u_2, v_2) = (0, 0))$, or,
- A moving object becomes ambiguous with another moving target; for example, Targets 1 and 2 move in such a way that they mimic each other.

Equation (12.5) can be simplified by taking the square on both sides of the equation. The equation can be written in terms of η as

$$\begin{aligned} &\left[((V_a - u_2)^2 + v_2^2) - ((V_a - u_1)^2 + v_1^2)\right]\eta^2 \\ &-2\left[x_2(V_a - u_2) - y_0(v_2 - v_1) - x_1(V_a - u_1) + y_1 v_1 - y_2 v_2\right]\eta \\ &+\left[(x_2^2 + (y_0 + y_2)^2) - (x_1^2 + (y_0 + y_1)^2)\right] = 0 \end{aligned} \tag{12.6}$$

Chapman, Hawes, and Nord assert that an approximate analytic solution can be found by defining d to be the path distance to the origin and is written as

$$d \equiv \sqrt{(V_a\eta)^2 + y_0^2 + h^2} \tag{12.7}$$

With this definition, the path length to Target 1 can be written as a Taylor series expansion about d:

$$\begin{aligned} d_1 \approx d + \frac{1}{2d}\Big[&(u_1^2 - 2u_1V_a + v_1^2)\eta^2 \\ &- 2x_1(V_a - u_1)\eta + 2v_1(y_0 + y_1)\eta \\ &+ x_1^2 + y_1^2 + 2y_0y_1\Big] \end{aligned} \tag{12.8}$$

The path length d_2 is of the same form:

$$\begin{aligned} d_2 \approx d + \frac{1}{2d}\Big[&(u_2^2 - 2u_2V_a + v_2^2)\eta^2 \\ &- 2x_1(V_a - u_2)\eta + 2v_2(y_0 + y_2)\eta \\ &+ x_2^2 + y_2^2 + 2y_0y_2\Big] \end{aligned} \tag{12.9}$$

For one target to "degenerate" to another, whether either target is stationary or moving, the path difference $d_{\text{diff}} = d_2 - d_1$ must be zero for all time during the imaging. This yield the following approximation:

$$\begin{aligned} d_{\text{diff}} \approx \frac{1}{2d}\Big[&(((V_a - u_2)^2 + v_2^2) - ((V_a - u_1)^2 + v_1^2))\eta^2 \\ &- 2(x_2(V_a - u_2) - y_0(v_2 - v_1) - x_1(V_a - u_1) + y_1v_1 - y_2v_2)\eta \\ &+ (x_2^2 + (y_0 + y_2)^2) - (x_1^2 + (y_0 + y_1)^2)\Big] \end{aligned} \tag{12.10}$$

With this defined, we can now explore the space where d_{diff} might be zero for all time during the dwell. So, let's start with the case where a moving target degenerates to a stationary target. Let Target 1 be the stationary target and Target 2 be in motion. This means that $u_1 = v_1 = 0$. The question then is: are there any values of u_2, v_2, x_1, y_1, x_2, and y_2 so that $d_{\text{diff}} = 0$? If all three terms of Equation (12.10) are zero for all of η, then this condition is satisfied. Let's take each of the terms in Equation (12.10) in turn.

The first term in the equation—the quadratic term—is a circle in (u_2, v_2) space. This can be seen by setting the term to zero:

$$(((V_a - u_2)^2 + v_2^2) - ((V_a - u_1)^2 + v_1^2)) = 0 \tag{12.11}$$

Recall that $u_1 = v_1 = 0$, so this can be written as

$$(((V_a - u_2)^2 + v_2^2) - V_a^2) = 0 \tag{12.12}$$

This yields

$$((V_a - u_2)^2 + v_2^2) = V_a^2 \tag{12.13}$$

This circle has a center at $(V_a, 0)$ and a radius of V_a. So, for d_{diff}, the values of u_2 and v_2 must lie on this circle.

We can do the same for the second term—the linear term—in Equation (12.10). Doing so provides the following expression:

$$x_1 = x_2 - \frac{u_2 x_2 + y_0 v_2 + y_2 v_2}{V_a} \tag{12.14}$$

Finally, we can set the last term—the constant term—in Equation (12.10) to zero. This yields

$$y_1 = -y_0 \pm \sqrt{x_1^2 + (y_0 + y_2)^2 - x_2^2} \tag{12.15}$$

For d_{diff} to be zero for all time η, there values of u_2 and v_2 for the moving Target 2 that lie on the circle defined by Equation (12.13). Furthermore, there are values of x_1 and y_1 [dependent on the relationships defined in Equations (12.14) and (12.15)] that satisfy Equation (12.6). All of these conditions point to the path lengths d_1 and d_2 being equal, which means that Target 1 will look like a stationary target at (x_2, y_2). Chapman, Hawes, and Nord assert because they are using the exact equation for Equation (12.6) that the solution is exact; that is, they assume no approximations to arrive at Equation (12.6). The approximations taken to get to Equation (12.10) point only the form that the solution for u_2, v_2, x_1, y_1, x_2, and y_2 must take for the degeneracy to occur.

This can be seen graphically in Figure 12.4. The circle lies in the (u, v) space and has a center at $(V_a, 0)$ and a radius of V_a. The stationary target, (x_1, y_1) has zero velocity in either the u or v, so it is at the origin on this circle. Thus, there are an infinite number of solutions for (u_2, v_2) that, if values for x_1, y_1, x_2, and y_2 can be found that satisfy Equation (12.6), will degenerate and appear to be a stationary target at location (x_1, y_1) in the image. There is a caveat: since V_a is likely to be much larger than u_2 or v_2, there are limits to where the solutions might lie on the circle. This is shown to be the gray box in Figure 12.4. Despite this condition, there is no way to solve for these conditions without further information.

A similar analysis can be done for moving objects that degenerate to other moving objects. For a moving object to have the same displacement and smear as another moving

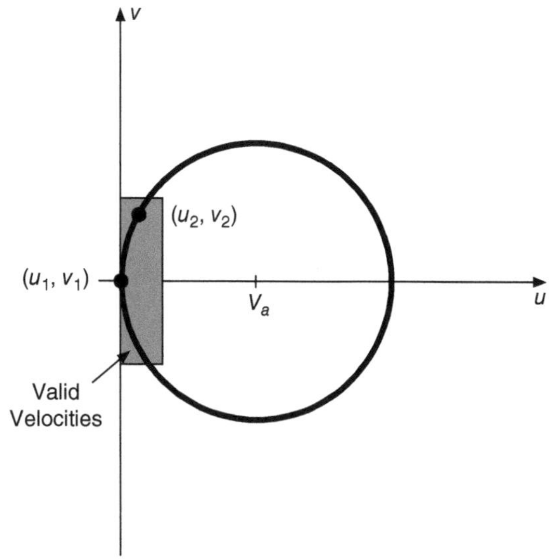

Figure 12.4 The family of solutions for (u_2, v_2) that could mimic a stationary target at (x_1, y_1).

object, they must live on the same degeneracy circle. This can be seen in Figure 12.5. Chapman, Hawes, and Nord show that (x_1, y_1) and (x_2, y_2) must lie on the same circle centered about the SAR in (x, y) space (i.e., they are in the same range bin). Thus, a moving object can appear like another moving object in the scene. This means that it is impossible to derive location and velocity for a moving target from the displacement and smear in

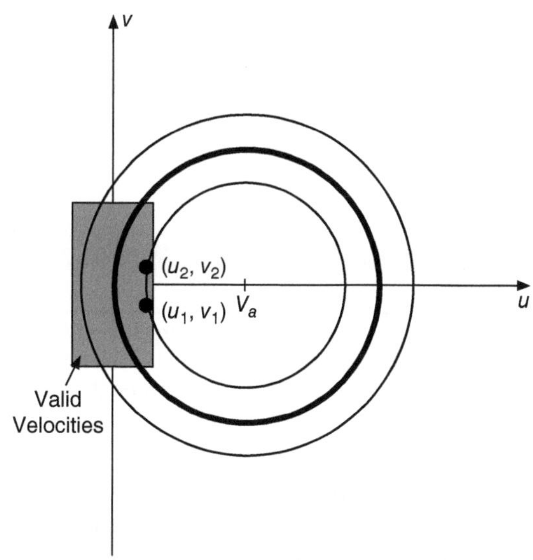

Figure 12.5 The family of solutions for (u_2, v_2) that could mimic a moving target at (x_1, y_1).

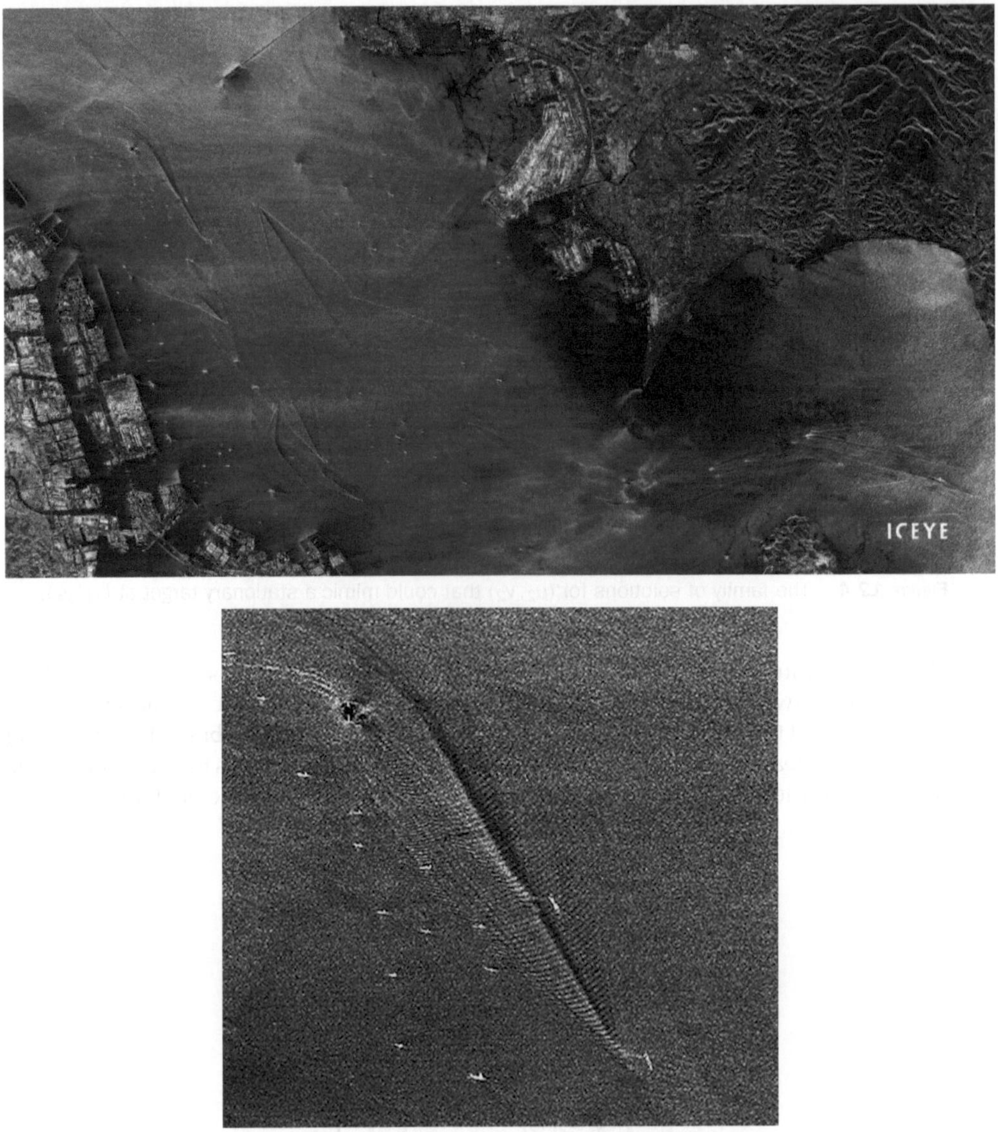

Figure 12.6 ICEYE image of Tokyo Bay, Japan. The range direction is along the rows of the image, while the azimuth direction is along the columns. The top image is the full image, while the bottom image is a zoom in on two maritime vessels heading into the bay (as seen on the left-hand side of the full image). *(Reprinted with permission.)*

a SAR image for a single-channel SAR system. Further information is needed to break the ambiguity. For example, in Figure 12.6, the ships are off their wakes. However, the information found in the wake (location and direction) can be used to find the velocity of the vessel. Additionally, if the Kevin wake is plainly visible in the image, information from that wake can be used to derive the composite velocity of the vessel.

12.3 Other Types of Moving Objects

Moving objects, such as ships, boats, trucks, and cars, are fairly common in SAR imagery taken of Earth's surface. Less common, but still interesting, are objects that move either rotationally[7] (such as a rotating marine antenna on a shipping vessel) or vibrationally[8] (such as powered by a rotational motor).

Rotating objects create interesting effects in the imagery. Rüegg, Meier, and Nüesch[9] studied and simulated how rotating objects manifest themselves in X-band stripmap SAR. They assumed very slow moving antennas with rotational frequencies of 0.5 to 1.0 Hz and objects between 0.75 and 25 meters in size. In essence, they found that rotating objects can create large displacement effects in azimuth. The displacement is highly dependent on the pulse repetition frequency (PRF). Keeping the PRF at 2 kHz, which would be common for a spacebased X-band SAR, the displacement for the rotational objects in azimuth could be as large as 3 kilometers. Reducing the PRF to 200 Hz reduced the displacement effect to 300 meters, making simulation feasible. Needless to say, in a high-PRF system, the displacement could wrap around the image along the azimuth axis, making visual detection difficult or impossible, particularly in a cluttered environment such as an urban scene.

Vibrational objects, such as a motor running in an idling vehicle, manifest themselves as paired echoes. Rüegg, Meier, and Nüesch[10] also studied vibrational objects in millimeter-wave SAR. They found that objects that vibrate at a certain frequency and a certain displacement (e.g., 30 Hz at 1 mm) will create paired echoes on either side of the mainlobe at the location of the object. The distance of the paired echoes is directly proportional to their vibrational frequency, while their signal amplitudes are directly proportional to the vibration amplitudes. Thus, the higher the frequency, the more spread out the paired echoes and the higher the amplitude, the brighter the paired echoes. In cases where the amplitude of the vibration is significant (e.g., 10 mm), multiple paired echoes appear on either side of the main object, like harmonics around a sinusoid. Rüegg, Meier, and Nüesch found that analysis of the moving objects with a time-frequency analysis tool, such as the Wigner-Ville distribution (WVD), could be very helpful. In particular, the sinusoidal nature of the vibrating object is clearly visible in the time-frequency representation.

---

[7] S. D. Silverstein and C. E. Hawkins, "Synthetic Aperture Radar Image Signatures of Rotating Objects," in Conference Record of the Thirty-Eighth Asilomar Conference on Signals, Systems and Computers, 2004, Pacific Grove, CA, 2004, Vol. 2, pp. 1663–1667, doi: 10.1109/ACSSC.2004.1399441.

[8] V. C. Chen, "Analysis of Radar Micro-Doppler with Time-Frequency Transform," in Proceedings of the Tenth IEEE Workshop on Statistical Signal and Array Processing (Cat. No.00TH8496), Pocono Manor, PA, 2000, pp. 463–466, doi: 10.1109/SSAP.2000.870167.

[9] M. Rüegg, E. Meier, and D. Nüesch, "Constant Motion, Acceleration, Vibration, and Rotation of Objects in SAR Data," Bruges, Belgium, October 2005, p. 598005-1 – 598005-12, doi: 10.1117/12.626529.

[10] M. Rüegg, E. Meier, and D. Nüesch, "Vibration and Rotation in Millimeter-Wave SAR," *IEEE Transactions on Geoscience and Remote Sensing*, Vol. 45, No. 2, pp. 293–304, February 2007, doi: 10.1109/TGRS.2006.887025.

Acronyms

ADC	Analog-to-Digital Converter
AFRL	Air Force Research Laboratories
AGL	Above Ground Level
ARIA	Advanced Rapid Imaging and Analysis
ARP	Antenna Reference Point
ATR	Automatic Target Recognition
BPA	Backprojection Algorithm
CAT	Computer-Aided Tomography
CCD	Coherent Change Detection
CNN	Convolutional Neural Network
CPA	Closest Point of Approach
DARPA	Defense Advanced Research Project Agency
DC	Direct Current
DEM	Digital Elevation Model
DFT	Discrete Fourier Transform
DLR	Deutsches Zentrum für Luft- und Raumfahrt e.V. (German Aerospace Center)
ECEF	Earth-Centered-Earth-Fixed
EM	Electromagnetic
EO	Electro-optical
FFT	Fast Fourier Transform
FM	Frequency Modulated
GEO	Geocoded
GIS	Geographical Information System
GPS	Global Position Satellite
GPU	Graphical Processing Unit
GRL	Ground Reference Line
GRP	Ground Reference Point
IFSAR	Interferometric SAR
IID	Independent and Identically Distributed
IMU	Inertial Measurement Unit
INS	Inertial Navigation System
InSAR	Interferometric SAR
IPD	Interferometric Phase Difference
IPR	Impulse Response
IRW	Impulse Response Width
ISLR	Integrated Sidelobe Ratio
JAXA	Japan Aerospace Exploration Agency

JPL	Jet Propulsion Laboratory
LFM	Linear Frequency Modulated
LFMCW	Linear Frequency-Modulated Continuous Wave
LIDAR	Light Detection and Ranging
MD	Map Drift
MLE	Maximum Likelihood Estimation
MTI	Moving Target Indication
NASA	National Aeronautical and Space Administration
NOAA	National Oceanic and Atmospheric Administration
NSP	NanoSAR Pod
PFA	Polar Formatting Algorithm
PGA	Phase Gradient Autofocus
POSP	Principle of Stationary Phase
PRF	Pulse Repetition Frequency
PRI	Pulse Repetition Interval
PSLR	Peak Sidelobe Ratio
QPE	Quadratic Phase Error
RAR	Real Aperture Radar
RCM	Range Cell Migration
RCMC	Range Cell Migration Compensation
RDA	Range-Doppler Algorithm
RF	Radio Frequency
RMA	Range Migration Algorithm
RMS	Root Mean Square
RVP	Residual Video Phase
SAR	Synthetic Aperture Radar
SLC	Single-Look Complex
SNR	Signal-to-Noise Ratio
SRC	Secondary Range Compression
SRTM	Shuttle Radar Topography Mission
VCO	Voltage-Controlled Oscillator
VOIR	Venus Orbiting Imaging Radar
VPH	Video Phase History
WGS	World Geodetic System
WVD	Wigner-Ville Distribution
ω-K	Omega-K Algorithm

Variables

β	Antenna beamwidth [as an angle] (deg)
η	Slow time (s)
θ	Squint angle (deg or rad)
λ	Wavelength of radar (m)
τ	The round-trip time delay for an energy wave to leave the sensor, reach an arbitrary object, and echo back (s)
ψ	Grazing angle (deg or rad)
A_t	Amplitude of transmitted signal (V)
A_r	Amplitude of received signal (V)
B	Bandwidth of radar system
D	Diameter of scene (m)
f	Independent frequency variable
f_0	Center frequency of radar (Hz)
G	Gain of antenna (dB)
h	Reference signal for quadrature demodulation
I	In-phase channel of demodulation; real component of complex signal
K	Chirp rate (Hz/s)
L_a	Physical width of antenna in azimuth direction (m)
P_r	Received power (W)
P_t	Transmitted power (W)
Q	Quadrature-phase channel of demodulation (90° of out of phase); imaginary component of complex signal
R	Range to any given object in the scene (m)
s	In general, the signal in frequency space
S	In general, the signal in time/image domain

$s_r(t)$	The real-valued, received signal before demodulation or dechirp
$s_dc(t)$	The complex-valued receive signal after dechirp
S_{mf}	Signal after matched filtering
$s_{\mathrm{qd}}(t)$	The complex-valued receive signal after quadrature demodulation
t	Fast time (s)
T	Pulse width (s)
v_a	The velocity of the sensor (m/s)
(x_a, y_a, z_a)	The antenna reference point at any given slow time (m)
(x_t, y_t, z_t)	A target reference point for an arbitrary target in the scene (m)
W_a	Width of antenna beam in azimuth direction (m)

Glossary

Active A system that uses its own energy to illuminate a scene

Aliasing An effect in digital signal processing where a signal whose frequency content is greater than that of the sampling frequency has the signal wrap around the frequency axis, providing a ghost response

Along-track A direction that is perpendicular to range

Antenna reference point (ARP) The point that is represented by the center of the antenna, defined here as (x_a, y_a, z_a)

Autofocus A technique that estimates phase errors across the aperture and removes them to improve image quality

Azimuth Same as *along-track*

Bandwidth The width between starting and stopping frequency in a linear frequency-modulated (LFM) wave pulse; inversely proportional to range resolution

Baseband The original frequency range (i.e., the bandwidth) of a transmission signal before it is modulated

Broadside An angle perpendicular to the flight track

Bulk compression Used in Omega-K, a reference function whose purpose is to compress the signal in both range and azimuth simultaneously; see also *reference function multiply*

Chirp rate The rate of change in frequency in a linear frequency-modulated (LFM) waveform

Closest approach The minimum point in range between the sensor and an arbitrary point in the scene

Coherent When two electromagnetic waves at a single frequency have a constant phase between them and that phase is known

Complex Numerical representation that contains two numbers (real and imaginary), which together represent magnitude and phase; also known as "imaginary numbers"

Compression Same as *focusing*

Cross-range Same as *azimuth*

Cross-track Same as *range*

Detected Term referring to the calculation of the magnitude or intensity from the complex signal (an image is "detected" for visual display)

Doppler centroid The average Doppler shift

Doppler chirp rate The rate of the slowly varying LFM chirp in slow time, also known as the azimuth chirp rate

Dwell The period of time in which the synthetic aperture radar illuminates a scene; analogous to term *slow time*

Effective pulse width The effective amount of time for which the LFM chirp lasts

Exposure time The amount of time a target is illuminated by the beam; for spotlight, this is equivalent to the dwell time

Far edge The edge of the swath furthest from the sensor

Fast time The time dimension, designated by the variable t, corresponding to the sampling of the radar returns from each transmitted pulse, typically measured in microseconds or milliseconds; it is directly related to range

Focusing The act of converting synthetic aperture radar data into a reflectivity map of the scene; see *image formation processing*

Grazing angle The angle created between the range vector and the ground plane, typically measured at the GRP, and, for spotlight SAR, in the middle of the imaging period

Ground reference line (GRL) The line passing through the GRP, parallel to the ground flight track; relevant for stripmap imaging, but not spotlight imaging

Ground reference point (GRP) The scene or swath center, typically defined as $(0, 0, 0)$

Image formation processing A series of algorithmic steps used to fully focus synthetic aperture radar data into a reflectivity map of the scene

Impulse response (IPR) The signal corresponding to a given object in the scene after compression, typically in the form of a sinc squared function

Incoherent In this context, when the phase of a electromagnetic wave is unknown or unmeasurable

Layover A geometric effect in SAR where tall objects are flattened in the scene because the range returns from the top to down of the object reaches the radar before the returns from the ground; the layover object points to the direction of the radar; can create odd visual effects in imagery

Mainlobe The part of the impulse response (IPR) that corresponds directly with the target; this is typically of the form of a sinc squared function and is the center part of the function

Map drift An algorithmic autofocus technique that can estimate and remove polynomic phase errors

Matched filter an optimal filter, implemented in either hardware or software, that can "match" a specific signal and produce a detection for that signal; maximizes signal-to-noise ratio (SNR)

Near edge The edge of the swath closest to the sensor

One-sided Fourier transform A Fourier transform that considers the positive or negative frequencies only; the amplitude is twice that of the full Fourier transform

Passive A system that uses the energy from an outside source, such as the Sun, to illuminate a scene

Phase error An unwanted phase term in the radar signal and it goes through processing; the most common phase error is quadratic phase error (QPE)

Projection slice theorem A slice of a 2D or 3D imaging scene where all of the information for the scene has been projected onto that specific slice

Pulse compression A signal processing technique, executed in either hardware or software, that will detect a linear frequency-modulated waveform and produce a signal impulse that corresponds to the distance of the object to the sensor

Pulse repetition frequency (PRF) The frequency of transmitted linear frequency-modulated chirps during imaging; see also *pulse repetition interval (PRI)*

Pulse repetition interval (PRI) The time interval between each transmitted chirp, including the chirp duration

Quadratic phase error A common phase error that takes on a quadratic polynomial form; typically due to improperly characterized motion from the imaging platform; typically removed either in motion compensation and/or autofocus

Range The direction along which distance is measured in the radar

Range-azimuth Also known as image space, the two-dimensional space where all data has been compressed in both range and azimuth

Range-Doppler Two-dimensional signal space where each pulse has been compressed in range, but not in azimuth/Doppler $S_{rd}(f, \eta)$

Ranging The ability to measure range from one point to another

Range migration The shifting of a signal from a given scatterer from one range cell or bin to another over the course of the dwell

Range skew From dechirp processing, a delay in scatterers from across the swath not lining up in time after the demodulation

Raw data The collected data before quadrature demodulation or dechirp

Real aperture radar Imaging radar where the beam is pulled across the ground and each azimuth resolution cell is made up of one or more pulses in range; resolution is determined by radar bandwidth in range and radar beamwidth in azimuth

Reference function multiply Used in Omega-K, a reference function whose purpose is to compress the signal in both range and azimuth simultaneously; see also *bulk compression*

Reference signal The signal used to demodulate the received radar signal; in quadrature demodulation, this is a sinusoid at the carrier frequency; in dechirp, this is a time-delayed version of the transmitted LFM chirp pulse

Residual video phase (RVP) From dechirp or stretch processing, a leftover quadratic phase error that, if the aperture is wide enough, must be compensated else image quality will be compromised

Scan Imaging mode where the antenna beam pattern is both dragged and slewed on the ground; it achieves a higher resolution than stripmap while gathering more area than spotlight

Scatterer An idealized point on the ground that scatterers the energy from the radar

Scene The area on the ground being illuminated by the antenna beam pattern over the course of the imaging period

Sidelobes The ringing on either side of the mainlobe for an object's impulse response (IPR) after compression

Slant plane The plane that extends from the flight track to a parallel track at any given range line in the scene; targets are projected onto this plane during imaging

Slow time The time dimension, designated by the variable η, that corresponds to the collection of fast time sampled pulses over the course of the imaging period, typically measured in seconds (or tenths of seconds); it is directed related to azimuth

Spotlight Imaging mode where antenna beam is slewed to a specific spot on the ground

Spread spectrum A technique used for transmitting radio, telecommunications, or radar signals; the practice of spreading the transmitted signal to occupy the frequency spectrum available for transmission

Squint angle The angle between the antenna reference point and the flight track; see also *broadside*

Stolt mapping Used in Omega-K, an interpolation technique to remove the range cell migration

Stripmap Imaging mode where antenna beam pattern is dragged across the ground

Subaperturing Act of subdividing the phase history into smaller chunks, leveraging the holographic nature of SAR

Synthetic aperture An artificial aperture created by the motion of the aircraft or spacecraft as a substitute for an impractically large real aperture

Target Same as *scatterer*

Unfocused phase data The phase data after quadrature demodulation or dechirp, but before image formation

Video bandwidth The reduced bandwidth after dechirp

Video phase history The data over the entire dwell after demodulation and analog-to-digital sampling

Zero-Doppler plane The plane created by the zero-Doppler vector, the nadir vector extending from the sensor to the ground, and the slant plane

Zero-Doppler vector The vector perpendicular to the nadir point at the flight track projected on the ground

Index

Note: Page numbers followed by *f* denote figures; by *t*, tables; by *n*, footnotes.

9 781260 458961

Printed by Libri Plureos GmbH in Hamburg,
Germany